T0202723

Science in Design

Textile Institute Professional Publications

Series Editor
The Textile Institute

For more information about this series, please visit: www.crcpress.com/Textile-Institute-Professional-Publications/book-series/TIPP

Science in Design

Solidifying Design with Science and Technology

by
Tarun Grover and Mugdha Thareja

CRC Press
Taylor & Francis Group
Boca Raton London New York

CRC Press is an imprint of the
Taylor & Francis Group, an **informa** business

First edition published 2021
by CRC Press
6000 Broken Sound Parkway NW, Suite 300, Boca Raton, FL 33487-2742

and by CRC Press
2 Park Square, Milton Park, Abingdon, Oxon, OX14 4RN

© 2021 Taylor & Francis Group, LLC

CRC Press is an imprint of Taylor & Francis Group, LLC

Library of Congress Cataloging-in-Publication Data

Names: Grover, Tarun, author. | Thareja, Mugdha, author.
Title: Science in design : solidifying design with science and technology / by Tarun Grover and Mugdha Thareja.
Description: First edition. | Boca Raton, FL : CRC Press/Taylor & Francis Group, LLC, [2021] | Includes bibliographical references and index. | Summary: "This book reveals the significance of the essential yet understudied intersection of design and scientific academic research and encompasses technological development, scientific principles, and the point of overlap between science and design. It benefits scientists, technologists, and engineers, as well as designers and professionals, across a variety of industries dealing with scientific analysis of design research methodology, design lifecycle, and problem solving"-- Provided by publisher.
Identifiers: LCCN 2020043635 (print) | LCCN 2020043636 (ebook) | ISBN 9780367548742 (paperback) | ISBN 9780367558000 (hbk) | ISBN 9781003095217 (ebook)
Subjects: LCSH: Engineering design.
Classification: LCC TA174 .G76 2021 (print) | LCC TA174 (ebook) | DDC 620/.0042--dc23
LC record available at https://lccn.loc.gov/2020043635
LC ebook record available at https://lccn.loc.gov/2020043636

ISBN: 978-0-367-55800-0 (hbk)
ISBN: 978-0-367-54874-2 (pbk)
ISBN: 978-1-003-09521-7 (ebk)

Typeset in Times
by Deanta Global Publishing Services, Chennai, India

Contents

Preface

What does society expect from technological innovations? How can they be calibrated with scientific techniques to guarantee their goodness?

These queries have always risen during the authors' research and teaching of courses on design and engineering offered to graduate and postgraduate students at different academic institutions. Today, the majority of the courses in academia focus on educating learners on a plethora of design disciplines, technological and engineering practices, and research methods to bring an innovative idea to life. But the value of design knowledge and methods in the improvement of the framework or the technology is only partially realized; the fact is that the importance of the scientific approach, while very important, is not extensively dealt with by educators and industry practitioners. Some recent studies show a dire need for design research in education that the academia sector is lagging behind, eventually creating a gap between knowledge gathering and its application for designing in practice. The contribution of design and technology research is always sought after during studies in engineering/design/management, or when defining objectives for upcoming research initiatives or during the instructional design of a course study. The need for this book became obvious when both authors initiated a search for proof of concept that addresses the aforesaid issues.

The book starts with the recognition of design as art, science, process, and outcome. A design as art holds ideation, and as a process illustrates the relationship between framework and development; equally important, however, science validates the selection and application of a theory or method for the discovery of knowledge, and as an outcome holds the benchmark for a new innovation. Design science as the science of design holds certain objectives toward the construction of different artifacts for a specific solution based on the environment, technology, and people involved in the domain problem space. Research in design science, as defined in this book, is unlike researching in the domain of social science or applied science.

The research questions involve whether the adopted artifacts are indeed the most efficient and viable as a means to reduce the complexity of a problem. Knowledge about the most advanced technologies and research methodologies; their degree of success in the contextual implementation of a solution; the scope of developing a novel creation, design innovation, or a computational algorithm: all of these should be considered as an agenda to operationalize hypotheses and design practices in order to meet the final criteria of effective performance in different dimensions.

Science in design reveals the design of a product or process aimed at accommodating a development cycle that is scientific rather than descriptive. The goal is to show how empirical research and the design cycle rely on a scientific approach and computational intelligence for validation of their purpose and the selection of design attributes. In our conception, design science research is a foundational practice that happens to be interdisciplinary and instrumental to the formulation and evaluation of various design theories and implementation in a given situation – to be validated by domain professionals in terms of relevance to the aim of a research project. The objective of this edition is to provide a comprehensive study of design knowledge and how it can be created using design science research and methodology, keeping in view technology, methods and practices, and knowledge utilization in a particular system's environment. Science "in" design is used here in a very specific sense: to denote domain studies and systems that have obtained a given behavior because they are adapted in particular reference to well-executed empirical methods. The artifacts and their construction in terms of behavior are driven by scientific explorations.

The environment of the physical world lies in the realm of natural science, but linking humans to the digital world of information and applications is possible in the domain of "the artificial". When an artificial system adapts to the behavior of humans in a controlled environment, it signifies the design of robotics, which becomes the objective of system development. The breakthrough technology of sixth sense robotics using gesture recognition has been included in this edition. Developing in some detail of two specific examples – design thinking and design science research – describes the shape of designers as they emerge from technological developments over the decades.

Beyond specific illustrations, the authors have indicated how the presence of science and technology is relevant to textile and apparel production, computing, graphics, fashion, etc. to all fields that converge on the essence of design to achieve specific goals and functions. Ultimately, this book will help you navigate through the complex layers of design concepts, background theories, design methods, and technologies for designing in different fields while providing you information on how to effectively think about using all these technologies and the appropriate research methodology to design the next.

We hope you enjoy learning from this book as much as we enjoyed writing it for you!

MATLAB® is a registered trademark of The MathWorks, Inc. For product information, please contact:

The MathWorks, Inc.
3 Apple Hill Drive
Natick, MA 01760-2098 USA
Tel: 508 647 7000
Fax: 508-647-7001
E-mail: info@mathworks.com
Web: www.mathworks.com

Acknowledgments

Writing the acknowledgments is an opportunity to pay gratitude to all those who matter in our lives and helped us along in achieving our dream and aspirations. With the grace of God and our efforts, we have reached this stage.

Owing this achievement to her paternal uncle and world-renowned author Dr. B. L. Thareja, Mugdha with her authoring partner Tarun appreciate the writings of recognized research authors like Dr. Reema Thareja, Dr. Shakti Kumar, Dr. Kaleshnath Chatterjee, Dr. Rajkishore Nayak, Prof (Dr.) Sanjay Gupta, and Dr. Asimananda Khandual, who have been a great source of motivation for us to step up in this journey toward contribution to research and education by writing simplified content focusing on transformative technologies and trends of the digital age.

Thanks to the two great leaders and philosophers whose inspirations helped us through the development of thought leadership: Dr. Vishwanath Karad, a well-known Educationist and Ambassador and torch-bearer of World Peace, and Dr. Mangesh Karad (Executive President, MIT Art Design & Technology University Pune) for providing a multidisciplinary university environment that nurtures aspiring minds and faculties with collective wisdom and collective knowledge of art, science, design, and technology to achieve the satisfaction of realizing impossible dreams. Fortunately, we have had the support of our fellow members of The Technological Institute of Textile & Sciences, Bhiwani Haryana, Mr. T. H. Ansari (Arahne Software), Mr. Sunil Kumar Puri (Techknit Overseas Pvt. Ltd.), Team Tukatech, and the Institute of Design at MIT ADT University, Pune Maharashtra.

Our special thanks to our family members – Mr. Jagdish Grover, Mrs. Kamlesh Grover, Mrs. Sukhvarsha, Mr. Mayank Thareja, Mr. Vipin Pahwa, Mrs. Harsh Pahwa, nephews, nieces, and other relatives – who were a source of abiding inspiration and divine blessings in letting us realize this dream.

Our heartfelt gratitude to Rebecca Unsworth and Helen Rowe of The Textile Institute, Manchester, UK, for being a pathfinder for us to initiate for the book proposal with the recognized publisher CRC Press, Taylor & Francis. Finally, we would like to thank the editorial team at CRC Press, including Allison Shatkin and Gabrielle Vernachio, for their sincere help and support.

Comments and suggestions for the improvement of the book are welcome.

I wish to dedicate this book to my late father Mr. Rakesh Thareja.

*Dad! As you look down from heaven, I hope
you're proud of your daughter.*

-Mugdha Thareja

Author Bio

Tarun Grover is currently working as Assistant Professor at MIT-ADT University, India, in the department of fashion design. A postgraduate (M-Tech) in Textile Chemistry with specialization in material studies and textile, Mr. Grover has worked with various universities of repute, technical research organizations, and government agencies for numerous projects. He is an alumnus of TIT&S Bhiwani, an institution established in 1943, renowned for its state-of-the-art research initiatives and higher education in Textile Science and Engineering. He wrote the Chapter "Selecting Garment Accessories, Trims, and Closures" in the book *Garment Manufacturing Technology* (1st ed.), published by Woodhead Publishing, and another chapter, "Integrating Sustainable Strategies in Fashion Design by Detox 2020 Plan – Case Studies from Different Brands" in the book *'Detox Fashion'* – (Vol.4) published by Springer Singapore. He is a lifetime associate member of TAI (Textile Association of India), and also a member of WDO (World Design Organization, Canada).

Mugdha Thareja is a research enthusiast and qualified professional in the field of computer science and information technology. A postgraduate (M-Tech) in Computer Science Engineering with a specialization in software engineering and image processing, Mugdha Thareja has around seven years of combined experience in academic research and education and the IT sector. She has served various premier engineering and higher education institutions in India as an academic faculty and has done research and publishing projects on the multidisciplinary approach. She possesses niche experience holding responsible positions at IT companies of global repute, including Genpact and e Software Solutions. She has offered her services as a consultant and instructional designer for various government projects sponsored by NABARD, TSSC, and NSDC in the field of technical education, skill development, and women empowerment. Her specialized areas of interest include software engineering, research methodologies, design research, agile methodology, human–computer interaction, user experience and multimedia, web technologies, and software project management.

1 General Aspects of Science, Design, and Engineering

1.1 WHAT IS DESIGN?

A design is a strategy or specification for the development of a system or object. In other words, design can be thought of as the implementation of a plan or a process to achieve an objective. More often, the term "design" is used to indicate the outcome of a process or strategy in the form of a prototype or a final product. The broader meaning of design can be interpreted more easily with the following narration.

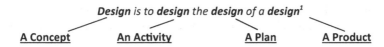

Undoubtedly, the term "design" encompasses the varying viewpoints of people who understand its meaning through different lenses. In general, "design" is the process of visualizing and planning the development of systems, objects, methods, products, etc. The definition of design is versatile and artful, and design has taken many shapes and forms based on multidisciplinary concepts. Besides the major cluster of design disciplines (engineering, information systems, industrial, architecture, textile, and fashion), experience design, service design, and interactive design have evolved as new domains for design enthusiasts to explore deeper insights.

1.1.1 DEFINITION OF DESIGN

First, let's glance at some of the noteworthy definitions of design by well-known authors.

A design is a plan to make something new for people that they perceive as beneficial.

– Koss Looijesteijn

Design is a reflective conversation with the materials of a design solution.

– Donald A. Schön

It's the difference between your favourite and least favourite thing you use.

– Scott Berkum

Design is its own culture of inquiry and action.

– Harold Nelson, Erik Stolterman

Having these in mind, the ability to design and the notion of design relating its associative aspects are the key focus in conceiving the definition of the term "design". Design can be classified based on three fundamental perspectives: art, problem solving, and the pursuit of the ideal [2].

a)   Ability to Design

The importance of designing ability has been realized by many design educators and practitioners for enhancing learners' skills to design. However, there is no clear picture of what constitutes traits of designing aptitude, since the various meaning of design makes this perspective a much broader concept. The classification of designing ability is aimed at encouraging efforts toward addressing challenges like contributing to building the future world, solving difficult problems pertaining to the environment, and acting as a guide for the next generation to follow. Therefore, the classification of design and designers' perspectives plays an important role in formulating a definition of design.

b)   Notion of Design

Here, we explain the three notions of design and discuss their influence on the development of a new product or system.

Class A: Art

Design is widely assumed as the expression of ideas in the form of sketches or drawings, commonly known as art. This is the classification of design on the basis of its common use. Although art seems to be creative, the creativity itself involves the process of transforming an imaginary picture into a concrete object. In particular, the process of creative art involves an examination of the past, since the image for an abstract idea comes from the designer's own mind and memory.

Class B: Problem solving

In this notion of design, the main focus is on how to design in the context of the current problem rather than on what to design. In this case, the procedural design is depicted within the framework of problem solving. Within this framework, the process of designing a solution requires identifying a problem and examining the gap between the existing state and the target design goal. In other words, the solution to a problem lies hidden in the designing gap. Therefore, a problem-solving activity cannot generate a new goal unless the procedure of determining the desired goal is complete. "To design is to plan, to order, to relate and to control", says Emil Ruder.

Class C: Pursuit of the Ideal

The term "design" can be used to explain the pursuit of certain ideals, i.e., solving evident problems. This notion of design can be understood easily from a social perspective and contains within it the definition of the future. The classification of design in the context of ideal pursuing refers to anticipating the future. It involves the process of abstraction in an ideal environment. Furthermore, this notion describes the nature of design that conforms to the future perspective that only humans can perceive. According to the design definition by Herbert Simon, "To design is to devise courses of action aimed at changing existing situations into preferred ones".

On the basis of above classifications, design can be defined as "the composition of a desired goal toward the future".

1.1.2 THEMES OF DESIGN

In order to set the theme of design, lots of different permutations and combinations can be applied, but the most straightforward way of explaining design would be through design education. Education in design is a way to know what and how, which are the means to a qualified design profession, which is a means to well-designed processes and products, which are means to economic competitiveness, which is a means to job creation, which is a means to economic wealth, which is a means to the quality of life [1]. There are a variety of approaches to defining the curriculum of design where the perspective, context, training, and goals may vary. Some of the approaches common to all perspectives are illustrated below (Figure 1.1).

*   *Project-based Learning*
    This is one of the best approaches to learn "design by doing design", which gives hands-on experience to learners and students. Throughout the project, the designer has the responsibility to define the problem, ideate and present designs, and then make refinements on receiving feedback through "Critiques".
*   *Visualization*
    With a foundation in aesthetics, designers practice visualization techniques to quickly sketch the abstract image and efficiently turn complex ideas or problems into easy-to-understand visuals.
*   *Insight, Research, and Co-creation*
    Designers need to walk in the shoes of customers to gain better insight into their problems. There are various methods to collect information or data about the defined/unknown problems. Examples include learning interview techniques, design research, co-creation, and mapping techniques, through which designers can grow their empathy for users and train to move past their own preconceptions and biases.
*   *Prototyping*
    Prototyping allows designers to rapidly build a test module, and then evaluate and iterate the design process based on the new concepts and feedback, saving time and money during a project.

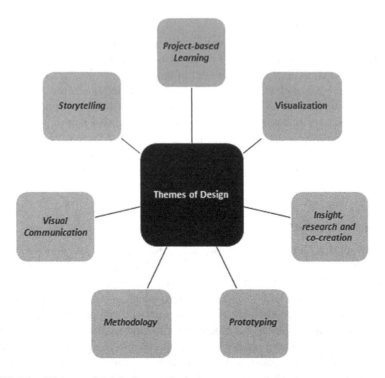

FIGURE 1.1:   Themes of the design curriculum.

- *Methodology*
  Methodology entails the final process for narrowing down to the best-fit solution after executing a number of trials that generate numerous concepts for exploration.
- *Visual Communication*
  Visual communication enables human power to receive and respond to visual information by seeing the shape, line, color, and type of visual elements. Having a strong visual communication theme allows designers to create different gestures and emotions.
- *Storytelling*
  We humans respond best to stories — it is how we naturally process and store information. Likewise, designers learn to harness this approach both in text and in user flows in order to sell concepts.

1.1.3   MULTIPLE FACETS OF DESIGN PARADIGM

According to Bryan Lawson and Kees Dorst (author of *Design Expertise*), "One of the difficulties in understanding design, is its multifaceted nature. There is no one single way of looking at the design that captures the 'essence' without missing some other salient aspects". Therefore, it is a big challenge to define the design in one frame. This is considered as a process of design activities – inputs, evaluations, and outputs. Design is truly multifaceted; it is a composition of many different disciplines of work coming together. The concept of the design paradigm is used to illustrate the model of an object that alters the layout with an aim to address changing demands and problems facing the distinguished design professions. A design paradigm can be envisioned as a prototype or a solution considered by a community as being effective, influential, and sustainable for growth. A design paradigm is a three-dimensional structure of a working relationship between groups of components and serves as an epitome for business success that ensures the quality of deliverables. The latest emerging and most powerful aspects of design paradigm are as follows:

- *Experience Design*: Experience design or XD is a cross-disciplinary perspective of design paradigm that involves the practice of designing an object, process, service, or environment, with a focus on delivering the best quality experience to the end user through interactive solutions. Experience design takes into account human factors applied during the design process. An innovative experience design rivets minds and increases the business appetite for radical ideas.
- *Eco Design*: Also known as sustainable design, eco-design involves the creation and redesign of products and services to bring configuration changes that reap many business benefits including but not limited to reduced cost, increased customer satisfaction, improved economy, and reduced damage to the environment.
- *Service Design*: Service design is the activity in which the designer specifies and creates a process to deliver an optimal solution for specific user requirements. It is a process of planning the arrangement of service components in order to increase the customer's interaction with the brand. It primarily informs the need for change in the existing process and for building a new product based on past experience design outcomes.
- *Sustainable Design*: Sustainable design is the practice of multiple disciplines that integrate an environmentally friendly approach and consider nature's resources as part of the design assets so that they exist in harmony with natural systems. The goal of sustainable design is to achieve a better future for the human race through the wise and low-volume consumption of Earth's resources. This design approach is most sought after in the concept of circular economy business models.

1.1.4 Design Questions

Questions are an interesting aspect of defining design, as questions are a concrete outcome of the everyday problems design tends to solve — including the core values and ideology of a designer. These questions do not require special tools and can be asked by anyone from any field, not the designers only. Having this, the design is inevitable, as long as there is a continuous stream of questions being asked with different objectives. Some of the most common questions that aid in formulating design definitions are the following:

- Who are the target customers? What is their common problem space? How to get the customer requirements?
- How do people communicate with and use the existing solution? What are its limitations right now?
- What are designers aiming to achieve? How does this impact the user experience?
- What is the objective of product design? How can it reach the people with a quick time to market?
- How can designers perform research and gather input from domain experts?
- What dependencies or regulations may affect the design experience?

After receiving answers to these questions and understanding the essence of new design and improvement, designers and research practitioners can definitely work on deriving the meaningful sense of design: what the "design" thing is.

1.2   FIELDS OF DESIGN DISCIPLINE

The term "design" has many synonyms, where each designer has a unique way to formulate the solution of any problem, whether it's related to the technical domain of design or to the applied arts. Underpinning the role of the designer in various fields might be different, but the principles remain the same, that is, to redefine how problems are approached, to identify opportunities for action, and to help deliver more complete and resilient solutions.

In the era of the 21st century, designers, researchers, and practitioners have been expanding the design to be more "designerly" through deep rooting in an interdisciplinary approach and methodological, conceptual, and theoretical frameworks to encompass ever-wider disciplines, activities, and practice. The following are disciplines which have been recognized as discrete design disciplines such as product, graphic/visual communication, interior, textile, retail and e-commerce, art and jewelry, furniture, and fashion design. The key foundational concepts that are recognized to have revolutionized the modern design arena are design management [14], design education, design research, and design science methodologies. These concepts play an important role across all the disciplines of design for the development of user-centered or humanized products (Figure 1.2).

1.2.1   PRODUCT AND INDUSTRIAL DESIGN

Industrial designers are also known as product designers, as they are particularly concerned with those aspects of products that relate to human usage and behavior and product appeal. In the view of manufacturing and marketing the products, design is the most important "feature" that helps designers grasp a real edge over competitors. Product design is the process of identifying a market opportunity, defining the problem, developing a proper solution for that problem, and validating the solution

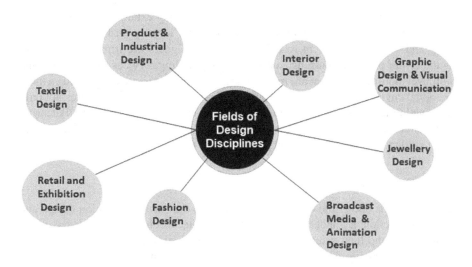

FIGURE 1.2:   Fields of design disciplines.

with real users. Therefore, designers have to follow a design process which starts from doing ideation through drawings, rough sketches, and illustrations of products, selecting the best match model, and then creating prototypes to demonstrate and test products. It is common for industrial designers as well as product designers to work as a part of product development team.

1.2.2   GRAPHIC DESIGN AND VISUAL COMMUNICATION

Graphic designing has always been the most striking benchmark of any successful brand in the market with an aim to increase business outreach. Various materials and software tools are being used for creating a brand identity that could enable communication with a broad audience across the globe. Both manual creativity using paper or printed materials and technology using digital software help in keeping in touch with customers and keeping record of feedback for further reproduction. Graphic designers develop and prepare information for publication with a sound understanding of text-based communication along with the smart utilization of communication properties, including symbols, colors, and pictures. The key role of graphic designers is to develop concept layouts and subcontract diagrams, illustrations, photography, and mock-ups to discuss project details with clients. The rapidly developing areas of digital media relating to the Internet and multimedia business presentations are opening up new areas of employment for graphic designers.

1.2.3   INTERIOR DESIGN

In the midst of urbanization, the migration of people to other cities in search of bread and butter as well as better lifestyles has increased requirements for qualified interior designers and design engineers in the real estate and architectural sectors. They are engaged in planning commercial and residential building exteriors and interiors by understanding the concept of minimalism, which is already pushing our style of living and work environment to the next level. Designers are trained in postulating the layout of space planning, space creation, building services, furnishings, and multilevel modifications that are more aesthetically consummate and provide a more functional environment. Today, the outlook of interior design plan requires not only a captivating look but also a designer to understand technical intricacies like structures, purchase of materials, risk and budgeting, furnishings, and most importantly expert skills to implement a project in good order to arrive at an optimal design.

1.2.4 Textile Design

Textile design is one of the disciplines that nurture the beauty of fashion design and apparel production. By understanding material science through the knowledge of textile materials and their scientific properties, designers in the fashion industry move forward to producing different categorized garments using the appropriate manufacturing process. The role of textile designers is not limited to the development of patterns, knit and weave construction, textures, and illustrations for fabrics, but also includes evaluating how the fabric appears and performs to meet the objective. Textile design as a discipline builds textile designers who are employed in the capacities of designers, trend analysts, dyers, colorists, and stylists in different market verticals like textile mills, export houses, design studios, merchandising business, craft houses, and retail stores. These designers are trained professionals having sound technical knowledge of each and every aspect of fiber to fabric processing including yarn making, weaving, knitting, dyeing, and finishing processes. Also, they demonstrate good command over the interpretation of different types of looms, knitting machines, and printing processes, which are in-demand skillsets of designers.

1.2.5 Retail and Exhibition Design

Retail and exhibition designers create and balance the building and installation of trade exhibitions, permanent shop displays, museum exhibits, pop-up stores, and interpretive displays. They use their inherent creative skills and those gained from the study of graphics, product design, and market research to capture, inform, and involve common people as the subject of their research and inspiration. The retail and exhibition design field is open to trained newcomers and professionals from industrial design, product design, graphic design, interior design, social science, and digital design areas like multimedia, web design, and e-commerce. Industrial designers use their abilities to design form and mechanical detail, interior designers use their abilities to design the function and aesthetics of spaces for human occupation, and graphic designers use their skills in presenting messages in visual form.

1.2.6 Fashion Design

In general terms, fashion design is considered one of the happening branches of design which attracts interesting characters, and its glamorous attributes tend to connect everyone. But designers have to really work hard to launch a particular collection inspired by a particular theme, which includes illustrations, drawings, pattern making, stitching, and an overall understanding of colors and fabrics.

Fashion designing has become very popular in India in the last few years, and many people are now considering it as a career choice. With economic prosperity, the average income of Indians has increased. Hence, people have better lifestyles compared to the past and they can afford to spend more on their attire. As a result, there is an increased demand for professionals who can design and create new garments, dresses, and attire to cater to the tastes of people from different walks of life.

1.2.7 JEWELRY DESIGN

The boom in the gem and jewelry industry has brought innumerable employment opportunities for jewelry designers. The role of designers is to conceptualize, prototype, and detail for manufacture items of jewelry such as rings, brooches, bracelets, necklaces, watches, eyewear, and earrings. They have specialized knowledge of the metals, jewels, precious stones, and other materials associated with personal adornment. They may develop designs for mass or batch production, or they may develop special items to satisfy one-off commissions. Jewelry designers may be employed within manufacturing companies specializing in jewelry or in other decorative personal and home wares such as silverware, cutlery, eyewear, watches, and trophies.

1.2.8 BROADCAST MEDIA AND ANIMATION DESIGN

A broadcast designer is a person involved with creating animated designs, graphic designs, and electronic media incorporated in television channels and production houses. The job of the designer is to create a look and feel for a specific idea or subject. Animation design is the art of creating special effects and other designs for various forms of media, including video games, movies, and even social media posts. The role of animation designers is to make animated TV shows or games, design settings, and create computer-generated images (CGI) to bring special effects to life. The industry of Broadcast Media and Animation Design has generated rapid growth in the unending demand for talented and aspiring individuals in this field.

1.3 SCIENCE AND DESIGN

Design as a field occupies more space when coupled complementary to science than any other field [3]. It was difficult to talk about any coexistence of science with design until the 20th century, when designers and science experts worked on various artistic monuments and the art of design proved its potential to bring life into objects. The creation of famous museum exhibits, innovative design of informative graphic material, or the interactive simulation of an experiment: all these can convey scientific insights in an intelligent, informative, and delightful way. Until the 19th century it was apparently believed that design relates closer to the fine arts than to science in general. Since scientific intellects ignored the commotions of the design space, designers and artists tended to share the same universities to learn skills and still share the same mindset.

In the recent course of history, the role of science in design has dramatically evolved with the introduction of new principles and practices. Earlier, scientific knowledge was related to direct experience, but now with the gradual development of different kinds of experimental theories and hypothesis, scientific knowledge has become fundamental and abstract and need not be associated with direct life experience only. Undoubtedly, scientific knowledge is more important in shaping the backbone of a design structure and needs the third eye of the user to realize its existence in the state of design. Scientific methods and scientific knowledge are increasingly used to develop a sophisticated design enabled with technology. For example, in a racing car in which the use of scientific knowledge provides the means for designing the system and operating its speed parameters, technology and mechanics provide power to the engine. To design scientifically is of paramount importance; this process is accompanied by methods like experimental variation, quantification, simulation, and mathematical descriptions. The social dimensions of scientific knowledge are useful in understanding the relations between science and social practices.

Science takes advantage of design research and development. The principles of empirical science rely on the observation of empirical data and are governed by empirical laws. Advanced scientific rules are based on scientific queries and research, intended to bring change into the world, improve quality of life, and create new horizons of knowledge. It is hard to decide whether science or design is more powerful in the modern technology era, but nonetheless they can benefit each other.

Bringing things to a head, this section introduces a special strand of relationship between science and design – Design Science, a scientific study of design artifacts that focuses on its multidisciplinary nature and recommends developmental research methods.

1.3.1 DESIGN SCIENCE

Design science is composed of two principal words – design and science. Where design depicts the sphere of multifaceted art, science underlines the logical patterns of the intent behind the selection of a particular design. The concept of design science was introduced in the late 1950s by R. Buckminster Fuller, who defined it as "a systematic form of designing". This term was later expanded in conceptual breadth by S. A. Gregory, who proposed a distinction between scientific method and design method.

According to Van Aken, the main goal of design science is to develop knowledge that professionals of different disciplines can use to design versatile solutions for real-world problems. Having said this and taking into account the wonders of design science research seen during the past few years, there is no denying the fact that the interactive collaboration happening between these two worlds offers a more focused scientific study of designed artifacts with the explicit intention of improving the aesthetics and functional performance of a system. Over an arbitrary period of time, the interchanging use of design science as the study of design and systematic designing has comingled the two meanings to a point where this term exhibits pragmatic behavior of both design as a science and the science of design.

1.3.2 DESIGN AND DESIGN SCIENCE

Nowadays, science is making a buzz by being applied across many fields, including wireless, mobile communication, automobile, healthcare, life science, information systems, defense, design, and engineering. Undoubtedly, design is the frontier of any product, which attracts its users' attention and compels them to experience the quality and utility of the product at least once. The common factor in both terms is the focus on the development of novel artifacts or the investigation of original artifacts that differ from existing ones. However, the objectives of both terms in terms of people and practices differ in their respective contributions to the knowledge base and in their ability to be generalized.

Design is a process to articulate a solution for a problem which is encountered locally and focuses on a particular target, organization, or end user, whereas design science works more toward the general interest of solutions designed to produce results that are relevant for the local and global community of practitioners and researchers. To reduce the gap between design as a practice and design science, the foundational requirements of design science research need to be satisfied per the following criteria:

1. Firstly, the purpose of creating new knowledge of general interest requires design science projects to make use of rigorous research methods.
2. Secondly, the knowledge produced has to be related to an already existing knowledge base in order to ensure that the proposed results are both well founded and original.
3. Thirdly, the new results should be communicated to both practitioners and researchers.

To understand these three requirements, consider an example of a project for designing a new electronic health record system. In order to classify the project as a subject of design science, the following three conditions in reference to the above criteria are postulated:

- *Use of rigorous research methods*: The foremost requirement of project development is to build an overall research methodology that encompasses problem investigation and the collection of relevant data, which can be carried out through survey questionnaires for large groups of healthcare professionals and expert interviews with physicians and healthcare providers working in relevant functional areas. Following this, the pre-evaluation of the artifact produced as a result of the analysis of collected data should be executed using grounded theories, research strategies, and methods.
- At the second stage, data inputs and the need for projected artifacts are analyzed in comparison with the already existing knowledge base of health

informatics and electronic information systems to point out the scope of improvement in the new version. Furthermore, this also helps in assessing the utility, originality, and validity of the artifact in context.

- The next step in the project is the dissemination of research outcomes and experimental results to both researchers and healthcare professionals through media like publication in journals and conferences, presentations at healthcare seminars, professional conferences, symposiums, and other similar events.

1.3.3 PRINCIPLES AND RECOMMENDATIONS

Current design research and methodologies advocate for design science as a present-day, state-of-the-art form of design beyond the abstract forms of human, natural, and social science. The core objective of design science is to create artifacts that are used in solving real-time problems [4, 5]. The parameters of design science are derived to meet certain criteria of rigor and relevance that approach a determination of their functionality in organizations' context of work, usefulness, and ease of use [7]. Design science aims at bridging the gap by bringing theory into practice, which is, however, not readily applicable for designing management processes. The primary goal of design science research is to manage planning, design, implementation, improvement, monitoring, or evaluation of information systems in an organization.

Design Science can be emphasized as an organized approach to design, making the design process itself a scientific activity. In general, the design process follows design science principles to create purposeful artifacts for the problem identified in design and management. The design science process is underpinned by several key principles, which we have summarized in Table 1.1.

1.4 EXAMINING DESIGN PROCESS FROM DESIGN SCIENCE OUTLOOK

Design Process is an approach for breaking down a large project into manageable and achievable chunks. Architects, engineers, scientists, and other thinkers use the design process to solve a variety of problems. A generic design process model [6] defines a designer's journey to tackle a project. Beginning with defining a problem, justifying the tasks, and generating prototypes followed by the design refinement, impact evaluation, and communication, a design lifecycle comprises of multiple small set of activities backed by background research. Nevertheless, gathering feedback from as many people as possible helps in making the decision whether to take the designed solution back through the process for improvement in quality standards.

Design science research is a practical approach for creating new artifacts to solve defined problems or for redesigning an existing solution to achieve the goal of process improvement. Two basic activities, namely "build" and "evaluate", are involved, where building is the process of designing an artifact for a specific purpose, and evaluation is the process of determining how well the artifact performs and fits into

TABLE 1.1
Principles of Design Science and Literature

Principle	Description	Key References
Design as an Artifact	The development of useful artifacts is a core requirement. Artifacts include: constructs to describe problems or solution components; models to represent the problem and its solution space; methods to provide guidelines for task performance; instantiations to demonstrate the utility of the artifact.	[4, 5, 8–10]
Design Problem Relevance	DS research is problem driven, aimed at addressing the problems situated at the intersection of people, organizations, and information technology.	[4, 8, 10]
Design Cycle	Design cycle activities iterate between building and evaluating artifacts and are based on both relevance and rigor, focused on addressing application domain requirements, while drawing on existing theoretical foundations and methodologies in the knowledge base.	[4, 8, 11]
Design Research Rigor	A design requires the use of methods and analysis appropriate to the tasks at hand. The DS rigor cycle links build and evaluate activities with existing foundational theories, frameworks, artifacts, processes, methodologies, and application domain expertise in the knowledge base.	[4, 8, 11, 12]
Design Artifact Evaluation	Rigorous evaluation methods are required to demonstrate the design of artifact's utility, quality, and efficacy. Metrics are used in comparing the performance of artifacts. Evaluation approaches may include case studies, field studies, analytical methods, experimental methods, testing, or descriptive methods.	[4, 5, 8, 9, 11, 13]
Design Research Contributions	Contributions of DS research include: an artifact that adds to the existing knowledge base; design construction knowledge improving foundations; design evaluation knowledge enhancing methodologies; experience gained from the design and evaluation of activities.	[4, 5]
Communication and Dissemination of Research Outputs	The results of design science research should be communicated and presented in an appropriate form to the technical and managerial community.	[4, 12]

the design framework [9]. In particular, a process-based approach follows multidisciplinary aspects of design and is evaluated in the following key seven design science principles, which are also listed in Table 1.1.

1. **Design as an Artifact**

   This principle demands that design science research focuses on building and evaluating an artifact in the form of a construct, a model, or a method [4, 5]. The process-based approach aims at producing a management process and produces therefore an artifact, which is a method to achieve something.

2. **Design Problem Relevance**

   The designed artifact should be relevant to the domain discipline [4, 8, 10]. According to Hevner et al. [4], artifacts in IS, for example, can be technology-based, organization-based, or people-based artifacts, which are all necessary to address problems in IS. On the other hand, a management process is an organization-based artifact.

3. **Design Cycle**

   Design science is described as a Generate/Test cycle and is therefore an iterative process to find a solution for a problem [4, 11]. In the process-based approach, a process is repeatedly applied and refined until it solves the problem. In addition, designing an artifact necessitates knowledge in the application and solution domain.

4. **Design Research Rigor**

   Design research rigor has to be evaluated in the light of how well an artifact works and not by how well it can be explained why it works [8, 12]. This brings the applicability and the generalizability of the artifact into the center of focus, which is also the primary goal of the process-based approach. Applicability and generalizability are achieved by application of the designed process in a number of organizations in different industries.

5. **Design Artifact Evaluation**

   Evaluation of designed artifacts is crucial for a design science researcher in order to justify an artifact's relevance for practice based on the business environment. In general, the artifact evaluation method is based on the criteria of measuring the utility, quality, and feasibility of an artifact.

6. **Design Research Contributions**

   The major contributions possible out of process-based design science research [4, 5] are as follows: (a) addition of artifact to existing knowledge body, (b) design development knowledge improving the generic structure and enhancing methodologies, and (c) experience gained from "build and evaluate" activities.

7. **Communication and Dissemination of Research Outputs**

   An important part of design science research is the effective communication and dissemination of research results [4, 12]. Hevner et al. argue that research in design science has to "be presented both to technology-oriented as well as management-oriented audiences" [4].

1.5 CONTRIBUTION OF DESIGN SCIENCE IN BUILDING DESIGN

Design is inevitably a continuous process, and its knowledge is entangled in science and technology. While the role of science in design remains a topic of deliberation and deep understanding, the development of artifacts using advanced scientific rules has been acknowledged to bring transformation to the world of design innovation, by improving the efficiency of the system and by creating new horizons for digital agility in the design process. Therefore, some sort of annotation is established between design science and design, which clearly opens up entirely new avenues of human endeavor. While design follows professionally recognized methods and its results are typically compared to the state of the art, design science on the flip side provides design researchers with critical analysis of outcomes based on the scientific body of knowledge and technological tools for evaluation and testing.

Gregor and Hevner et al. [15] proposed that the contribution of design science research knowledge to the maturity of the design research process and its artifact construction can be explicated in two dimensions: Application domain and Solution. An application domain maturity graph seeks improvement in the practice for which the contribution is intended, whereas a solution maturity graph states the maturity of artifacts that can be utilized as a foundational step toward finding an appropriate solution. Based on these dimensions, a 2×2 matrix depicting the four coordinates of design science contribution framework is illustrated in Figure 1.3:

a) *Invention*: This is the act of bringing new solutions for new problems, leading to radical innovation. An invention that achieves a completely unique function or result may be a comprehensive breakthrough. As Jeff Bezos puts it, "Every new thing creates two new questions and two new opportunities". The sharing of ideas in networks with actors from various sectors unlocks the potential for innovation. The panoply of such works is novel, and their contribution can enable new practices and create the basis for new research fields. Some examples of inventions are the first car, the first X-ray machine, and the first data mining system. For digging out the way of inventions require a good understanding of existed work and patience with a bit of luck in order to occur.

b) *Improvement*: This kind of contribution brings new solutions for addressing already existing problems. The contribution of design science in improvements is to propose solutions that actually improve on state of the art in efficiency, usability, safety, maintainability, or other qualities of the product. Some examples of improvements are the first sportbike, an X-ray machine with substantially reduced radiation, and a data mining system able to handle very large data sets.

c) *Exaptation*: Exaptations occur frequently in design science research in order to adapt an existing solution for a problem for which it was not originally intended. In simple words, an existing artifact is repurposed, or exapted, to a new problem context. For example, Hydroxychloroquine drugs are used to fight against malaria but after thorough research and evaluation,

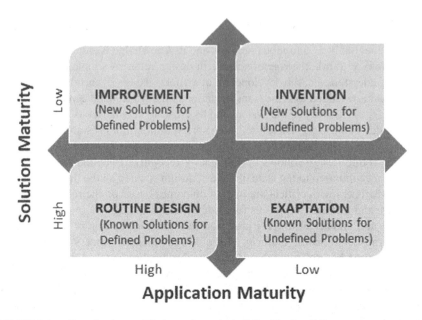

FIGURE 1.3: Contributions of design science in building design [15].

they have been acclaimed to save the lives of corona patients. To understand from the designer perspective, the designer usually produces not only an artifact in isolation but also a use plan for it. The use plan tells when and how people should use the artifact. Another example seen in the field of human-computer interaction is called reappropriation. Initially, cameras were built into mobile phones to capture images, but later on they were transformed into scanners, mirrors, note-taking tools, etc.

   d) *Routine Design*: This kind of contribution is determined for known solutions for known problems. This kind of credential is used for incremental innovation that addresses a well-known problem by making minor modifications to an existing solution. Much practical professional design would fit into this category, e.g., the design of a new smartphone with slightly better specifications than its predecessor. Routine designs typically do not count as design science contributions because they do not produce new knowledge of general interest, but they can still be valuable design contributions.

The following section details the multidisciplinary nature of science in design, along with relationships and practices in different disciplines.

1.6 PEOPLE AND PRACTICES

A set of different activities is carried out by people in an orderly way that produces some meaningful results. In general, this cluster of activities performed by people is known as practices. One example would be the practice of an orthopedic doctor who engages in repairing injuries to the musculoskeletal system and many other activities related to the spine and other bones and joints. When humans practice their skills, they utilize different tools to handle certain objects. An example is the use of X-ray machines and bone pillars for joint and knee surgery.

Practices can be categorized as formalized practices and structural practices. Some practices involve individual performing actions, while other activities involve the participation of groups. Engaging in some practices may give rise to some practical problems that need immediate attention. These practical problems can be puzzling questions, troublesome situations, or obstacles to achieving the desired result. The invention of the X-ray is one such innovation that offered aid to medical doctors dealing with the limited ability to view the deep structure of internal organs.

1.6.1 SCIENCE AND ENGINEERING PRACTICES

The strong relationship of science with engineering has played an important role in shaping the evolution of design. The important practices in science and engineering with underlined research pamper the development of scientific knowledge to develop model and carry out evaluation. The research cycle starts with asking questions, formulating problems, and continuously engaging learners in developing computational solutions to represent meaningful data.

The scientific body transforms knowledge into new ideas giving birth to new innovations. Engineers work on the principles of the foundation of science in application development to create systems in a controlled environment.

1.6.2 DESIGN AND ENGINEERING PRACTICES

Design with science can be described as a scientific study of emerging methodologies, tools, and practices. It is accepted across multiple domains that the blend of design and engineering has made technical innovations possible, affordable, and accessible to the wider community. The advent of engineering design has built a dynamic relationship between the science of design and design engineering. Redefining the way design methodology was assumed to be independent of the nature of problems and type of knowledge used, today the impression of the design process is validated based on analysis and scientific study of different engineering domains and their degree of involvement in design methodology.

The interlacing of design and engineering practices can be looked at in design journals having valuable contributions of authors from different engineering disciplines. A focused design process should also take the mechanical behavior of the product into consideration for a project to be successful. A product in a physical form (e.g., lever, elevator, footwear, tech textile) not only should be aesthetically appealing but also should be able to perform the underlying engineering function or mechanics it was designed for.

1.6.3 DESIGN SCIENCE AND PRACTICES

There is an intricate relationship between a design science project and its scope to local and global practices. In connection with design science with local practices, Figure 1.4 illustrates that design science projects may or may not utilize empirical data from that particular practice, but its results may contribute to building a scientific knowledge base. This body of knowledge, while benefiting local practitioners and the research community, should also be of significant relevance to a global practice.

Having this, a design science project seeking to contribute to the creation of a new knowledge base and to the global community can still be executed within the boundaries of local community practice. Thus, stating that design science is able to reach to a broader range of practices is not superlative, since the researcher remains involved in the design science project within the local community practice from an initial stage, building artifact solutions using the knowledge base, till he succeeds in generalizing the design of the artifact and refining the design knowledge toward the end of the project in order to obtain a generic solution for global practices as well.

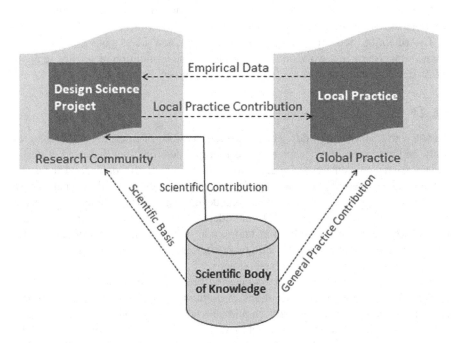

FIGURE 1.4:   Local and global practices in design science research [16].

1.6.4   SCIENCE AND DESIGN ENGINEERING

Right from the invention of simple devices like the wheel and the spoon to the complex models that designers create today, almost all artifacts involve design engineering with a scientific rationale behind it. The design of an artifact like a table fork as a product is available to the consumer in different shapes and materials, but the fundamental utility remains the same – to help eating. While the visual forms may differ, there is a common and essential component of engineering design that makes it useful to eat food. Not surprisingly, the foundation of any engineering design is based on the scientific approach, which implies the utility of scientific theories and principles, technical information, and pattern realization for a system to perform predefined functions at the highest optimal efficiency.

Nevertheless, a scientific basis is evidently present in the basic components of a design process, including the formulation of design objectives and evaluation criteria, creation, and testing. Primarily, design in engineering projects applies scientific knowledge to derive appropriate and valid solutions for technical problems. Understanding the importance of science in design engineering helps designers to ensure that the end product is made to achieve the purpose while using sophisticated design tools to automate the involved tasks systematically.

1.7   KREB'S CYCLE OF CREATIVITY: BLURRING THE BOUNDARIES

The ground truth knowledge of theories stating the degree of association between different disciplines comes from Neri Oxman's exemplary diagram of "Kreb's Cycle of Creativity". Featured in the *Journal of Design and Science* in 2016, Oxman's illustration (Figure 1.5) establishes a holistic view of the mutual relationship between four domains – Art, Science, Design, and Engineering.

Dividing the cycle into distinguished boundaries and dominions, the Kreb's Cycle [17] aimed at bringing together design and science in a way that further allows a deep understanding of the interrelationships among these disciplines.

According to Kreb's Cycle of Relationship, design and science lie opposite to one another in one circle. In contrast to engineering and design, science and design as individual disciplines do not depend on each other for input or output in the cycle. But they do form an incredible connection if we dig down deep by making a lens to view the fusion of design and science. Design and science are linked in an interactive yet complex relationship that exists with its own patterns of recognition.

> Science converts information into knowledge. Engineering converts knowledge into utility. Design converts utility into cultural behaviour in context. Art takes that cultural behaviour and questions our perception of the world.
>
> **– Neri Oxman**

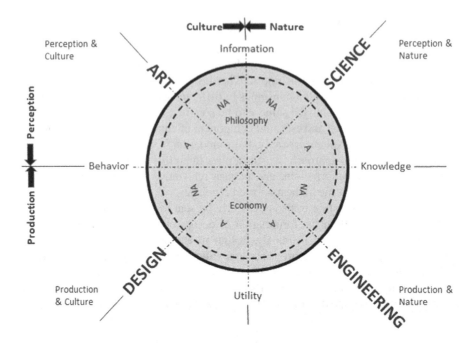

FIGURE 1.5:   Krebs cycle of creativity (A – applied, NA – not applied).

1.8   DESIGN THINKING – INNOVATING DESIGN WITH COGNITIVE SCIENCE

Design thinking is a universal approach for solving a set-in-stone range of problems by means of designing better solutions, services, and experiences. Design thinking as a panoply of design science outlines a completely new and dynamic approach to hitting the touchpoints of consumer behaviors and expectations. It is an integral part of innovation that allows consumers, teams, designers, and organizations to have a human-centered perspective with a scientific approach to address a problem. Design Thinking addresses a wide range of issues and is best used for bringing about innovation in multiple contexts [18].

- Redefining design value
- Human-centered innovation
- Quality of life
- Shifting markets and social behaviors
- Issues relating to corporate culture
- Issues relating to new technology
- Reinventing business models
- Scenarios involving multidisciplinary teams
- Entrepreneurial initiatives
- Educational advances

- Medical breakthroughs
- Problems that data can't solve

1.8.1   Nature of Design Thinking

Design thinking is in many ways the inverse of scientific thinking [18]. Where scientists explore theories and facts to discover insights, designers discover new patterns to address facts and possible solutions. In a world with increasing problems that greatly need attention to understanding the need for ideas, design thinking makes a valuable contribution toward innovative design strategy for creating a new solution.

Design thinking encourages the development of real collaboration between the designers and practitioners to solve emerging problems of different types. On the organizational front, adopting design thinking principles will help remove the existing silos and lubricate the system through team initiatives and cross-disciplinary research. Moreover, design thinking aims at creating a conductive environment for design innovation.

1.8.2   Design Thinking and Innovation

Design thinking in a process-based approach offers understanding capabilities as well as means for grappling with environmental factors to act immediately to the change in human behavior. This is crucial in developing and refining the methods used to create a working model of the final product.

FIGURE 1.6:   Design thinking and interactions [18].

In order to embrace design thinking in a real sense, and to innovate for sustainable development, we need to ensure that we have the right things in the right place and in order, e.g., mindsets, collaborative teams, and conducive environments.

Figure 1.6 illustrates that creating the right mindsets, selecting the appropriate team, and setting up environments for exemplary innovation to take place are three of the essential aspects of fostering sustainability in organizations of all scales.

To create new innovative solutions, design thinking helps to develop a truly open, explorative work culture and ethos and combines the magic of both analytical skills and imaginative piece of work. In view of this, famous technology kingpins like Microsoft, IBM, and Google invest in creating state-of-the-art workspace environments for employees where they can play with real-world objects, dissemble, rearrange, and increase their appetite for creative thinking through innovative solution designs. Some companies have adopted the policy of sending their entire staff on team-building getaways where they behave in their natural manner, arrange rafts together, jump around in circles and, in the best way possible, nurture their minds like kids. The goal is to make employees feel comfortable and safe and invent breakthrough solutions in a playful manner.

A purely technocentric view of innovation is less sustainable now than ever, and a management philosophy based only on selecting from existing strategies is likely to be overwhelmed by new developments at home or abroad. What we need are new choices – new products that balance the needs of individuals and of society as a whole; new ideas that tackle the global challenges of health, poverty, and education; new strategies that result in differences that matter and a sense of purpose that engages everyone affected by them. It is hard to imagine a time when the challenges we faced so vastly exceeded the creative resources we have brought to bear on them.

– Tim Brown

1.9 TOWARD DESIGN INNOVATION

There is a close relationship between design and innovation. Design is the backbone of innovation, and innovation is considered as the driving force of a market economy. The elaboration of design innovation as a new concept contributes to the academic research and industrial discourse of design. A conceptual definition of design innovation provides the basic tools for understanding design models to build innovation theory inspired by design, currently dominated by professional engineering discourses [20].

Design by innovation is understood and used in various conventions. In general, design innovation depicts the process of product development, where the resulting output is a complete product or a module alone. The importance of design innovation is recognized as a means of achieving excellence in niche product competition. The essence of design innovation is the need for the continuous improvement of products and the creation of new designs with ideal features (Figure 1.7).

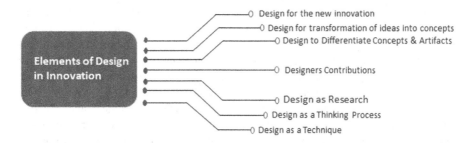

FIGURE 1.7:   Elements of design in innovation [22].

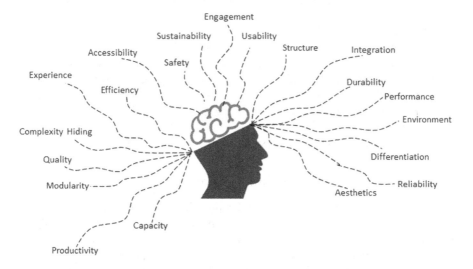

FIGURE 1.8:   Mind mapping of design innovation [24].

1.9.1   FEATURES OF DESIGN-DRIVEN INNOVATION

Innovation is defined as both a process and an outcome. It has been widely accepted as a driver of ongoing success in the competitive market. A growing number of research studies have a major focus on unfolding the translucent connection between design, innovation, and performance improvement. Design is increasingly recommended as a strategy rather than purely as a function within the innovation task. The process of innovation by design generates certain characteristics that are novel from various aspects, including components and their features, functional operations, scalability, and framework associated with the new design [21] (Figure 1.8).

1.9.2   DESIGN INNOVATION BY COMMUNICATION

The digital revolution has created new technological demands for multifaceted designers. In the current environment, it's no longer acceptable to have knowledge about design aesthetics only; one must also know about holistic product development. Designers working on innovation paradigms should be able to demonstrate their intelligence, coding knowledge, and practical exposure to building communication strategies based on user experience research. Communicating design concepts and ideas in an empathetic manner is far more likely to support a well-thought-out design process down the road.

Design-driven innovation explores the ways in which designers and research communities of art, science, and engineering are influencing each other toward creating new products for the materialist culture world. Assuming the result of design innovation requires bringing an imaginary design blueprint to life. Some general questions that contribute to problem formulation and method selection follow:

Q1: What makes a product great?
Q2: What is the need for designing a new product?
Q3: What is the role of design firms in the development of creative products?
Q4: How is the role of people and technology changing the facets of design innovation?
Q5: How are the methods of design communication creating benchmarks for innovation?
Q6: How does visualization of design inspire a design innovation?

1.9.3 Languages of Design Communication

Communication design is a system-based approach that focuses on the exchange of media and messages to communicate instantly with the people. Examples of this approach are used to create new media channels to ensure the message reaches the target audience along with already existing channels like print, crafted, electronic media, or presentations. Basically, this approach is designed to integrate multimedia components into a single model rather than a series of discrete efforts within the culture or organization. The following are a few languages of design communication that play a significant role in communicating a design to a larger audience [19].

1. **Visual Stories**

   To design a visually strong product, it must have an emotional component in the form of a story, since a story will leave permanent footprints on the mind of the user. The concept of visual design can be taken in a broader sense to understand that the designed entities we see, or otherwise perceive, can't exist without a bit of a story behind them. Many products in the market become obsolete and neglected by consumers due to bad visual storytelling that fails to stand out among various other competitors. To capture a good share in the market, designers have to come up with a strong identity not only for product or service but also for the customer and end users. A new design can be developed by utilizing the various elements of visual communication like color, position, texture, size, orientation, shape, tone, images, symbols, animation, and videos in order to deliver the intended message more effectively.

2. Use of Technical Terms

Another approach to reach the design target is by considering technical terms related to the product. This is purposefully done to build customers' trust in products by making them aware of the featured process used for designing the product, e.g., responsive design, immersive reality, etc. Technical terms depict the core environment, components, and behavior of a design system (Figure 1.9).

3. Analogy

Design-by-analogy is a powerful tool for design innovation, especially when thinking about conceptual ideation. The approach of analogous design thinking helps designers to identify design concepts and define their goals. Analogy in design communication is a cognitive process that is assumed to be a major source of new concepts that link cross-domain knowledge by means of common attributes and the relationship between the end user's situation and other related areas. One such tool is bio-mimicry, which applies learning from nature toward problem solving. Taking insights from natural systems like honeycomb pattern, designers can create analogous spaces and communicate their inspiration behind the design of structures.

4. Gesture

Gesture has been studied from various perspectives, sometimes with respect to computer support for human communication and collaboration but also with respect to the psychology of gesture (see chapter 6). Human gestures are used predominantly to depict a number of aspects of the design. The General use of gestures for design communication has been found in building HCI systems, robotics, interactive dialogue systems, collaborative task completion tools, semiotic analysis, and sign language development. Gesture recognition is an emerging approach that suggests the design of new technologies for interacting with digital information and involves the use of hand and arm movements.

5. Typography

Typography is all about adjusting the text within the design while creating powerful content. It provides an attractive appearance and preserves the aesthetic value of your content. It plays a vital role in setting the overall tone of your website and ensures a great user experience (Figure 1.10).

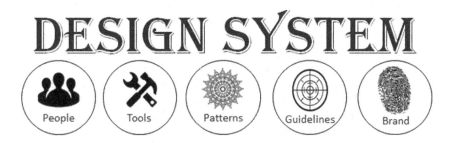

FIGURE 1.9: Components of a design system [23].

FIGURE 1.10:   Understanding of typography.

FIGURE 1.11:   Manual sketch and digital sketch.

The advent of science and technology has been continuously pushing designers to come out of their comfort zone and use digital tools to produce advanced sectional drawings. Designers use typography to communicate different feelings and create a mood that enhances the brand's influence.

6. **Sketching**

   Sketching offers ways to convey and receive messages about that which is tacit. Earlier, most designers simply relied on manual sketching, but now designs are entirely done with computer-aided design software that aids in drawing global orders. There is no doubt that sketching by hand allows a designer to capture an idea quickly, but due to its limitations is restricted in its ability to draw multiple sketches, and it is also difficult to sketch in a very short cycle. Therefore, CAD/CAM technology has brought a new hope in providing computer-based methods, which can provide three-dimensional renderings, rotating or moving the "sketch" in realistic ways on the screen and allowing a product to be assembled and disassembled in a virtual world. Thus, end users may observe and react to amazingly realistic representations of products still in the idea stage (Figure 1.11).

If a sketch in whatever form gives the end user an idea about functionality, it offers more than a hint of the design language to be applied. Several such sketches may offer more about what kind of "soul" the product will have than any well-phrased verbal description: emotions, messages, and meanings are tacit, not fully describable with words, formulas, or mathematics.

Design denotes thought processes and practices that result in the development of an object. Many roles that a design activity can play in innovation and how that activity leverages different design theories toward the success of innovative products are often studied to limited details. Different types of methods and techniques used by designers for creating products and services, independently or in collaboration, are also classified in terms of design. In fact, design indicates an association between research, ideation, and professional activity.

1.10 SUMMARY

This chapter has suggested the importance of science as a foundation for designing a product that contributes to the betterment of the modern world. The combination of design and science plays an important role in bringing out the quality deliverables that are inherent in design to the sciences. In order to ensure success in the science–design collaboration, both communities need to explore more and engage with each other. The sociological perspective allows us to better understand the peculiar nature of design and to reveal "why" and "how" the role of science is engaged with design, art, and engineering.

Furthermore, the latter sections described the assessment of a design process using the lens of design science principles and guidelines. As many of the existing guidelines for design science research (e.g. [1]) focus on artifacts, we found that the design process–based approach can be a very useful component that refines and adapts the design science approach to build and test management processes in system designing.

Future scope of literature review and hands-on research in the same field will be sure to test and modify the communication pathways of system units to meet the functional and non-functional needs of the system. This chapter encourages discussion on design science methodologies and pitches potential recommendations for the development of design processes.

REFERENCES

1. Hardt, Michael. *2006 The term design.* http://www.michael-hardt.com/
2. Taura, Toshiharu, Nagai, Yukari. *A Definition of Design and Its Creative Features* (2009)
3. Owen, Charles L. *Design Thinking: Notes on Its Nature and Use* (2006)
4. Hevner, A.R., March, S.T., Park, J., Ram, S. "Design science in information systems research". *MIS Quarterly* 28(1), (2004): 75–105
5. March, S.T., Smith, G.F. "Design and natural science research on information technology". *Decision Support Systems* 15(4), (1995): 251–266
6. Bonollo, Elivio, Montana-Hoyos, Carlos. "Drawings and the development of creativity and form language in product design" *ACUADS Conference* (2011)

7. Peffers, K., Tuunanen, T., Rothenberger, M.A., Chatterjee, S. "A design science research methodology for information systems research". *Journal of Management Information Systems* 24(3), (2007): 45–77

8. Hevner, A.R. "A three cycle view of design science research". *Scandinavian Journal of Information Systems* 19(2), (2007): 87–92

9. Iivari, J. "A paradigmatic analysis of information systems as a design science". *Scandinavian Journal of Information Systems* 19(5), (2007): 39–64

10. Baskerville, R. "What design science is not". *European Journal of Information Systems* 17(5), (2008): 441–443

11. Hevner, A., Chatterjee, S. "Design science research in information systems". *Design Research in Information Systems*, Integrated Series in Information Systems book series (ISIS), *22*, (2010): 9–22

12. Benbasat, I., Zmud, R.W. "Empirical research in information systems: The practice of relevance". *MIS Quarterly 23(1)*, (1999): 3–16

13. Pries-Heje, J., Baskerville, R.L. "The design theory nexus". *Management Information Systems Quarterly* 32(4), (2008): 7

14. Pandza, K., Thorpe, R. "Management as design, but what kind of design? An appraisal of the design science analogy for management". *British Journal of Management* 21(1), (2010): 171–186

15. Gregor, S., Hevner, A.R. "Positioning and presenting design science research for maximum impact". *MIS Quarterly* 37(2), (2013): 337–355

16. Goldkuhl, G. "From action research to practice research". *Australasian Journal of Information Systems* 17(2), (2013): 57–78.

17. https://spectrum.mit.edu/winter-2017/neri-oxmans-krebs-cycle-of-creativity/

18. Dam, Rikke Friis, Siang, Teo Yu. "Design thinking: New innovative thinking for new problems". https://www.interaction-design.org/literature/article/design-thinking-new-innovative-thinking-for-new-problems

19. Williams, Anthony, Cowdroy, Robert. "How designers communicate ideas to each other in design meetings". *At International Design Conference - Design 2002*, Dubrovnik (May 14–17, 2002)

20. Mutlu, Bilge. "Design innovation: Historical and theoretical perspectives on product innovation by design". *5th European Academy of Design Conference held in Barcelona* (2003)

21. Ardayfio, David D. "Principles and practices of design innovation". *Technological Forecasting and Social Change* 64(2/3), (2000): 155–169

22. Hernández, Ricardo J., Cooper, Rachel, Tether, Bruce, Murphy, Emma. "Design, the language of innovation: A review of the design studies literature". *She Ji: The Journal of Design, Economics, and Innovation* 4(3), (2018): 249–274

23. https://uxdesign.cc/everything-you-need-to-know-about-design-systems-54b109851969

24. https://simplicable.com/new/design-innovation

2 Design Science and Research Methodology

2.1 INTRODUCTION

Most recently, design has gained the attention of educationists and researchers as a well-recognized discipline. In design education, design knowledge exchange is instrumental in fostering creativity and enabling learners focus on scientific approaches to design as a forethought rather than an afterthought. Furthermore, design research benefits from the use of scientific methods for knowledge discovery and decision making. Design is often conceptualized as art, science, fashion, process, and product. While design is relatively a tenderfoot in the global open research community, most of the foundation theories studied in science, technology, and engineering fields describe the trans-disciplinary metaphor of design. In industry, design is used as a terminology to refer to a diverse range of innovative activities performed to achieve business success.

A multidisciplinary design approach looks toward integrating the methods and skills of designers from different disciplines into a joint effort. Multidisciplinary design professionals essentially understand how multiple areas of domain expertise can collaborate to solve practical design problems. Design knowledge embodied in research not only takes account of objects, environment, and systems, but also looks over their domain vocabulary, contexts, solutions, and consequences. As a point of focus, the presence of science in design knowledge always remains a cornerstone for creating the design ontology (Figure 2.1).

2.2 ONTOLOGY OF DESIGN

The concept of design ontology is majorly accepted across the areas of design and engineering. Enterprise-grade projects require a collection of systems and human resources to operate in a given situation. Sometimes, to meet the changing process requirements, a group of engineers having access to individual systems is subdivided into small teams to perform a particular set of tasks quickly. Undoubtedly, the degree of interaction between resources will determine the success of the product development project.

Design Ontology is a prominent framework that underpins efficient product development by the use of design knowledge, unified concept description, information exchange, and multicultural team understanding. To meet the need for standardization of domain terminologies, generalized design ontology is thought to be the foremost step in a research-driven design process that allows the reuse of domain knowledge in the dimensions of understanding, description, and explanation. The

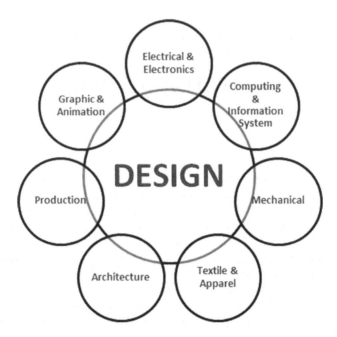

FIGURE 2.1: Design as a multidisciplinary concept.

methodology of building design ontology leverages a deep understanding of domain knowledge, design research, and design implementation. A unified space set of domain vocabulary and axioms serves the purpose of disambiguation using an approved design model.

2.2.1 ONTOLOGY AND DESIGN KNOWLEDGE REPRESENTATION

The representation of design knowledge [1] is a knowledge standard and technology to constitute design knowledge management. Design knowledge representation differs from information processing in a research project. Expression of knowledge implies understanding the content of the design knowledge, whereas information processing in general deals with the determination of the amount and format of information (Figure 2.2).

Let us assume a scenario where the working engineers use apparently identical signs to describe the domain concepts with different semantics and use these concept descriptions in different manners throughout the development process. Herein, a lack of coordination is identified because engineering designers bring diverse background knowledge to their respective job profiles.

Design disciplines have an enduring history of synthesizing their knowledge base through the process – the creation of artifacts followed by evaluation of their performance. Figure 2.3 shows the process cycle of knowledge management in design disciplines. Knowledge is acquired through the building and execution of actionable

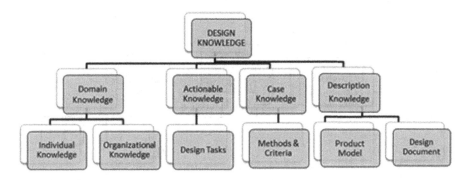

FIGURE 2.2: Types of design knowledge.

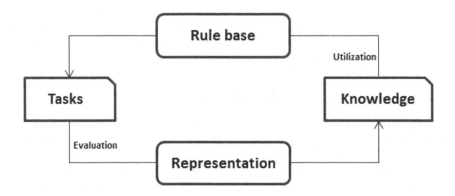

FIGURE 2.3: Knowledge creation and utilization.

tasks. The two blocks – rule base and representation – are the channels that govern the operations of a discipline (Figure 2.3).

2.2.2 SCIENCE AND DESIGN ONTOLOGY

The purpose of ontology-based knowledge representation is to acquire the design knowledge of related domains and provide a common understanding to design researchers, as well as to determine the common domain vocabulary providing concise definitions of constituent terms and their relationships [1].

The Science of design investigates the creation and integration of domain artifacts in context across different system environments. Driven by empirical research and computational science, design ontology sanctions the integration of derived domain models into logical frameworks. In design science, system and product development are addressed by encapsulating analysis and construction and deducing from relevant scientific disciplines (Figure 2.4).

Leveraging ontology-based knowledge [4,5] for designing has its own advantages such as interoperability, formalizing business processes, and effective methodology

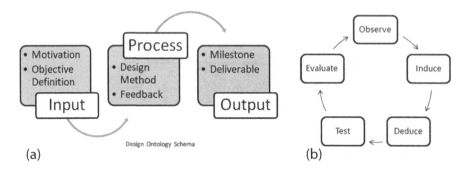

FIGURE 2.4:   (a) Design ontology schema; (b) empirical research.

implementation in knowledge management practices. Building a unified framework of design ontology gives formal knowledge description about the level of interaction between system and individual by means of domain knowledge, artifacts, database, and algorithm optimization.

2.3   ENGINEERING RESEARCH METHODOLOGY

The relationship between the research problem and methodology [27] is bidirectional. One might begin with the topic of research interest, select the appropriate research methodology, and then iterate the process to refine the hypothesis for further solution design. In engineering research, there have been many promising methodologies that exist to support the research process. But they are not rendered evenly in this diverse field of disciplinary research. The following methodologies, commonly termed as emerging methodologies, can be adopted in research activities across the STEM disciplines.

   a. Case Study
      Case study as a research methodology is not highly accepted by researchers due to its assumed limitations in many global regions. Not surprisingly, the experimental results obtained by a case study in one discipline are assumed to be of less importance to scientific development and to have less scope of generalization for other disciplinary research. However, if understood well, a case study methodology can help in investigating the success of solution design in response to technological and organizational issues.
   b. Discourse Study
      Discourse study or discourse analysis is a research approach based on linguistic modeling. The objects of discourse analysis are defined in terms of syntax, semantics, and parts of speech. Contrary to general linguistics, discourse analysis studies the use of language beyond the sentence boundary. In engineering research, discourse study focuses on concept negotiation to understand the phenomenon of project-based pedagogies.

c. Inductive Research

Many engineering problems require extensive research collaboration to find an appropriate solution. Inductive research is a systematic methodology in the field of engineering science that involves design and development of grounded theories using methodical data analysis. This research methodology starts with a research question and mandates researchers to analyze or review codes for base elements that have been extracted from the data. A large amount of data collection results in multiple iterations of the review process in which codes are grouped into categories. These categories become the source of building new theory. Thus, inductive research differs from conventional research methodology where the process of gathering data and its analysis are tightly interlaced, and it offers alternative conceptualizations of data for researchers to validate the use of adopted theory for solving the domain problem.

d. Action Research

Action research as a methodology practice reflects continuous improvement through the interlinked process of action and research in parallel. This type of research methodology is largely adopted by science and engineering practitioners to bring desirable change. Kurt defines action research as "a comparative research on the conditions and effects of various forms of social action and research leading to social action" that uses "a spiral of steps, each of which is composed of a circle of planning, action, and fact-finding about the result of the action".

2.4   DESIGN RESEARCH METHODOLOGY

The field of design studies has expanded its scope to become an interesting and important research discipline. Design research has evolved as a fast-paced field of exploration with its increased vitality in helping designers and researchers build products with improved quality and features that are useful for their socio-environmental benefits. Every field associated with design knowledge using process uses different typologies to build design research methods. These design research methods are implemented on three major pillars – communication, exploration, and implementation.

The communication method builds upon a deep understanding of the perceptions, needs, and challenges of people in their daily life as well as of domain experts in different functional areas who are engaged in providing services to their customers. Through this approach, a design researcher gets the benefit of acquiring knowledge to think of design on the basis of both customer and business perspectives. An inspiring design research method makes use of different tools at each stage as detailed below:

Communication

   i. *Interviews*: Interviews are planned conversations that are structured in a way to learn about challenges and issues faced by people, to get inspired by the way they solve problems, and to work in concord during their initial journey for better decision making. Another source of interviews is the expert who possesses a deep knowledge and industry experience in the area where the same cardinal problem is being solved with the new approach. As with expert interviews, the most inspiring individuals may get to participate in analogous industries.

ii. *Group Discussion*: Group discussions bring together a group of individuals from different walks of life. The biggest challenge in a group discussion is to come to know the individuals and to build empathy with people who feel ineffective in speaking to other vocal participants of the group. Formative evaluation is one of the ways to obtain the opinions and specific feedback of the participants to refine the design research plan.

Exploration

i. *Survey*: Research practitioners in the design discipline seek inspiration through surveys that offer a broader view of user experience, design needs, and industry demands. There is a plethora of commonly available tools and web portals like Google Forms, Microsoft Forms, and Survey Monkey that make it easy to share a questionnaire with an extended group of individuals. Surveys are used to quantify the observations across the groups and represent them in statistical or graphical form. Therefore, this calls researchers on to spend effort and time in creating questions that empathize with the target audience and capture information about each individual participant, leading to an effective survey. After launching the survey, explore the pattern of response by 15–20 people and find if some questions might have misinterpreted while answering inline. Surveying is an approach to prototyping the solution model and serves as a tool to reduce further risk.

ii. *Data Research*: An important part of exploration in the research process is gathering information from literature and the internet in order to create knowledge. Searching on Google inundates the result pages with a wealth of information that is relevant to some projects. In addition to research journals, there are links to relevant blog posts, project websites, scientific papers, and whitepapers, which are all useful to understand current problems and focus on the most challenging issues. Take time to analyze the gathered materials and make use of infographics to comprehend complex information. Good data research helps you shape survey questionnaires or interview questions in a better way and frame a versatile design model as you learn new things with this ongoing activity.

iii. *Data Mining*: Data mining is one of the prominent ways for finding inspiration from existing literature. Data science in the design process is the most important tool used by designers and industry experts today to accelerate their design value with faster deployment and a quick time to market. In order to extract knowledge from the resultant data, play around with intelligent visualization tools or spreadsheets to illustrate the pattern matching of interrogated data.

iv. *Experiments*: The selection of an experimental approach is determined by the scope of work by the researcher and is subject to the conditions and focus groups. The three experimental approaches used in exploration in design research are the following: pre-experiment, quasi-experiment, and true experiment. An experimental design research focuses on quantitative or statistical methods to validate the requirements and evaluate the adopted theory in practice.

v. *Analogous User Experience*: Putting yourself in the customer's shoes is about getting an empathetic experience. However, another vital part of design research process includes requirement analysis, using experience from the real world outside the desk, where similar problems have been solved using different techniques and design models. Analogous experience, for example, can be witnessed in the healthcare industry, which leverages technology and data exploration by learning similar user expectations from a study of the hospitality industry. Analogous experience looks for situational analysis to determine the design framework and is as simple as a customer buying products from the online marketplace.

Implementation

A design research methodology or DRM [3] is implemented as a set of supporting methods and as an approach in order to generate a relevant framework for design research. Design research activities involve the creation and evaluation of the adopted theory in support and the model of the desired situation. In general, the framework of DRM is comprised of four stages [2] [8]:

1. Research Clarification
2. Descriptive Study (I)
3. Prescriptive Study
4. Descriptive Study (II)

The vitality of the design research framework can be understood for a research project use case where the goal is to improve the approach used to execute the initial stages of a design process. This underpins the importance of task understanding by the design researchers when existing design support is considered ineffective for improving the success factors of a product in development. Hereby, the researchers do not focus on their initial idea but choose to exercise a structured research approach, conforming to the framework of DRM (Figure 2.5).

2.5 DESIGN SCIENCE – CONCEPT AND METHODS

The meaning of Design Science [11,12,25] draws out the concept of scientific design by including systematic knowledge of design methodology and design process, as well as underpinned scientific study of designing of the artifacts. The growing interest of

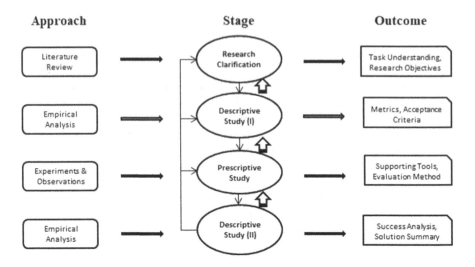

FIGURE 2.5: DRM framework.

the design community within information systems and technology in scientific information discovery highlights the parity between design science, design theory, and DRM. Generalizing different opinions of scientists and designers about the science of design and science in design, design research is regarded to have encompassed the meaning of these different terms in one constitution of Design Science Research (DSR). The following section offers a walkthrough of preconceptions and de facto descriptions regarding design science [6] knowingly acquired by researchers and professionals having a common cognitive interest in design foundation.

DESIGN SCIENCE – STUDY OF DESIGN

Design is central within the research and study domain, and, therefore, it is instrumental to the tasks of design science. But design science extends beyond design study alone. The study of design science is oriented toward comprehending and improving the construction of artifact with a goal to solve a problem. The top searched arena for design science, information systems, is one of the many problem-driven study fields including but not limited to management, electronics and communication, computer science, technology, and embedded systems.

DESIGN SCIENCE – THEORY OF DESIGN

Despite having precise elements that interlace the principles of design and product development, theories of design are distinguished from the study of natural science in the manner that design theories are driven by evaluation and prescribed metrics underpinning the design methods. Much of the literature presumes design science to be derivative of design theories, while some extracted literature throws light on design science as the mode of testing theories by designing objects to solve practical problems. The idea of the pre-eminence of design science depicts this as methodology and as a normative mode of scientific and design knowledge discovery, more than design theory alone.

DESIGN SCIENCE – IS IT ACTION RESEARCH?

The nature and essence of action research is often confused with the study of design science. However, both are focused on problem-solving with fundamental mathematical evaluation. However, it is not surprising that the approach for design science is not necessarily performing action research. Action research is fundamentally distinct from design science for the fact that design science underpins the construction of artifacts for solving a problem, whereas action research utilizes social studies and organizational behavior to solve a problem. In action research, the knowledge is discovered by taking measures. On the other hand, design science attempts to discover knowledge by designing stuff to overcome a challenge. Design science is not action research, but a paradigm for subject research methodology.

DESIGN SCIENCE – IS IT A METHOD?

Design science involves methodical ways for designing components. And thus, methods can be created for design science activities. But these methods are not particular to the study realm in which they are implemented. The domain study of design science involves the nature of science to create purpose-driven artifacts with adaptive behavior. As stated in the preceding paragraph and as a matter of fact, design science is not a method or methodology but more of a research paradigm.

DESIGN SCIENCE – IS IT A COMPUTER SCIENCE FIELD?

In the modern era, the understanding of design science as computer science stands void with the extension of design science activities ranging across many disciplines. Computer science is focused on designing computer-specific artifacts. However, the study of IT artifacts lies at the focal point of computer science, information systems, and design science.

Moreover, design science as a primary research paradigm has a much broader scope for artifact designing including fields like management, architecture, engineering, and economics. Design science for academics in management and technology provides a multidisciplinary description of fundamental methodologies. This explanation of research methods can be leveraged for deducing empirical research problem formulations common to information technology and information management.

DESIGN SCIENCE – IS IT AN AUTONOMOUS DISCIPLINE?

Design science extends beyond the boundaries of any particular academic discipline. The development of design science as a scholastic discipline is necessarily trans-domain with many educational fields involving the study of design methodology. Pinpointing its vital roles in education and research, design science is also assumed to be important for professional designers to discover scientific knowledge in the selection and building of artifacts. Design science is not a separate academic field but has its presence in many disciplines of humanities, engineering, science, technology, and management, which have design as the common subject of interest.

DESIGN SCIENCE IN REALITY

There are undoubtedly a number of things to learn about design science facts and myths. Enumerating some of the concepts confused with design science as described above can be helpful in developing an idea of what design science is. To commence with exploring the definitions of what design science is, and of design science as a research paradigm, the discussion of design research and design science in Robert Winter's article "Design Science Research in Europe" [7] takes first place as introduction.

To understand the essence of design science, we need to know the concept of scientific design, which refers to modernized design based on scientific knowledge, in contrast to craft-centric, pre-industrial design. The scientific design process utilizes a mix of inherent and derived design methods. The emergence of the concept of design science is deeply connected to the history of scientific designing that traces back to the periods of 1920 and 1960. A 60-year timeline and now the desire for scientific design process appears to be happening around again in the modern age of design revolution. In order to develop a new product, we need a design method with a scaled organization of finite functional objects. For illustration, planning the construction of a house in architecture designing, a series of definite patterns are sketched, engineered, and rendered to conclude the final product.

The reemergence of science in 21st-century novel design methods lay in the background application of computational and scientific methods for solving designing problems related to the engineering, technology, and management disciplines. Design scientists work on identifying the components of existing framework, enabling designers to build change configuration for the components of new framework. Nevertheless, design science deals with practical improving problems in contrast to wicked problems that lack application of scientific approach to design methods.

DESIGN SCIENCE METHODOLOGY

Design science [6, 17] involves building an artifact to improve product quality for stakeholders and empirical analysis of the performance of an artifact. Here, an artifact refers to the component of the design process for knowledge discovery including annotations, methods, algorithms, and techniques used in the projected system. The performance of artifact, in general, is investigated in the context of design, implementation, periodic maintenance, and reuse of system functionalities to achieve its goal.

The methodology of design science attempts to recognize the laws of design from the practical knowledge of the natural sciences, to visualize the selection of artifacts appropriate for designer's use, and to develop rules by formulating core design methods into scientific design methods. In a nutshell, the involvement of science in design methodology extends beyond the definition of scientific design and contains a methodical knowledge of design process and modus operandi, as well as the scientific reinforcement of designing artifacts. And therefore, the science of design (design science) can be understood as the alliance of subdisciplines comprising of design as their common cognitive trait.

2.6 DSR

DSR is an emerging novel research methodology [15] aimed at probing the design-oriented problem-solving theories and knowledge in different fields like science, technology, engineering, and business management. DSR methodology lays the fundamental logic of building an artifact that is intended to address an unsolved problem.

To solve any problem, research using design science [9] goes through the two cycles – design and empirical. In the design cycle, artifacts are designed to assist stakeholders. The empirical problem-solving cycle involves the extraction of intended answers in context by designing artifact knowledge. In other words, a problem in DSR is solved by construction and inspection of an artifact in context. For example, the cycle on one hand may involve design and investigation of architecture design techniques and, on other hand, may involve design and investigation of routing algorithms in networks.

> **Design Cycle:** In the design cycle, we iterate over examining the nature of the problem followed by treatment design and design validation. Different design problems may require different levels of insights and keen efforts to reach a decision.
>
> **Empirical Cycle:** The empirical cycle, similar to the design cycle, starts with the analysis of a research problem, design, and validation of design setup. Validation of the design setup is carried out to ensure the inferences in makeup are supported by the framework.

2.6.1 OUTCOMES OF DSR

March and Smith proposed four general forms of DSR output in their widely acknowledged paper in 1995, showing the contrast between DSR and general science research. These four generic forms of output in DSR termed as follows:

1) Construct: This outcome depicts the conceptual knowledge base of a problem domain. This is generated during problem formulation in the context of conceptualization and undergoes refinement throughout the design process.

2) Model: This is a set of statements representing the relationship between two or more constructs, i.e., how things should be and how they are actually showing up. The integrated model is designed on the basis of its application and theories describing relationship between two or more constructs.

3) Method: A method reflects the goal-oriented algorithm that is used to modify construct or perform a task so that the solution space model is recognized.

4) Instantiation: This output of DSR defines the realization of artifact by operationalizing other three outputs: construct, model, and method.

2.6.2 INDUSTRY OUTLINE OF DSR

1) *"Client as Designer in Collaborative Design Science Research Projects: What Does Social Science Design Theory Tell Us?"* – In this study, Weedman focused on the idea that artifact building is vital, as major problems are caused by disconnects between the customer world and the design world, and less by differentiating elements in multidisciplinary realms.

2) *"Design of Emerging Digital Services: A Taxonomy"* – In this paper, Williams, Chatterjee, and Rossi examined design taxonomy for the classification of leading digital services. This classification further implements two major classifying dimensions, design objective and service provider objective, respectively. This research contribution to analysis of the DSR framework can be utilized in the development and inspection of a fundamental system model with finite artifacts and model components.

3) *"Designing Enterprise Integration Solutions Effectively"* – This paper by Umapathy, Purao, and Barton sheds light on the study conducted on the reuse and effectiveness of design knowledge on design outcomes. The reuse was analyzed for the development of system integration solutions, out of pattern-based enterprise integration. The two major contributions of this paper toward DSR framework in industry have been (1) the construction of a methodical artifact that helps designers in building integration tools on the basis of pattern analysis and design strategies and (2) the evaluation of the artifact to validate its usefulness in meeting the design goals.

4) In his paper *"Essence: Facilitating Software Innovation"*, Aaen has proposed an approach to enable innovation in software development using agile methodology. Representing a problem-oriented transformation of common artifacts, this study illustrates the comparative analysis of design processes and components.

2.7 FRAMEWORK OF DSR PROCESS

The realm of design research has been engaged in exploring feasible approaches to combine design and research as well as to investigate the designing process. There have been interesting parallels while observing differentiators of design in information science and industrial design. In a broad view, both disciplines validate the design of systems to accomplish the requirements. The description of design science in the process of design research upholds the fact that design science is an explicit, yet organized and strategic approach to design, and not just the leveraging of scientific knowledge to build artifacts. Essentially, the role of science in design can't be negated since design in some sense is a scientific activity in itself.

2.7.1 ACTIVITIES

The framework of any disciplinary research is based primarily on the scientific approach and design activities. In its secondary dimension, it utilizes the produced outputs of the design research process. The complete set of activities can be divided into the following: identification of research problems and knowledge flows; selection and building of artifacts; and evaluation process. The generic process of design research emphasizes situation analysis, problem formulation, solution synthesis, and stakeholder communication. For every design project, ample methods are chosen for step-by-step activities that result in sizeable tracks throughout the design process.

Building upon the framework for design research on the top of scientific knowledge provides a deep view of the roadmap for design science researchers. In the DSR process, the description of concepts, artifacts, functions, and people roles ensures relevance to the research goals [9] and thus contributes to domain knowledge discovery. Approaching design science methods for research problem solving, the goal of research is projected at improving artifact design to increase products' scope of usage for different business areas and academic fields. To ensure that the promised improvement has been met, domain experts use the existing knowledge base and work on generalizing the solutions followed by rigorous evaluation (Figure 2.6).

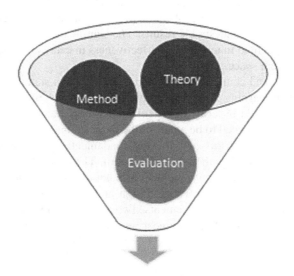

FIGURE 2.6:   Model view of DSR.

Modeling a DSR framework [13,16] encapsulates the activities of (1) theory foundation, (2) solution design, (3) action research evaluation, and (4) simulation. This section of the chapter will read these four activities and then look into a dimensional view of each.

(1) *Theory Foundation*: As noted in the article by Hevner [24], DSR is tipped off by business requirements and existing theory space. In the interest of having an informed design research process, theory building should be put into action as both an antecedent and a consequence of design research.

As a predecessor of solution design activity, theory building in DSR paves an approach to reduce problem complexity by means of utility theories and hypotheses. Distinct bare-bone forms of theory formulation can be implemented to analyze improvement and to contemplate the effectiveness of a solution (Figure 2.7).

The component of solution design depicts the concepts and assertions that constitute the technology implementation for a given problem. The component of problem formulation describes the individual's understanding of the problem that is being solved by the proposed technology, in context to the relationships with other aspects of problem space, i.e., the business requirements being addressed by the solution design.

A utility theory or hypothesis then connects the concept of technology in the solution design space to defined aspects of the problem space that is being addressed. A technology in solution design can be used for event compensation or reduction of a problem. By eliminating one or more factors causing a given problem, a solution technology helps in reduction. By treating the situation of a problem space, a solution technology attempts to compensate for undesirable results. The significance of theory building should be specified in terms of its effectiveness in catalyzing improvement over a problem space.

A precise and complete description of the problem statement is needed for effective learning about the problem space. A solution technology is generally assessed in terms of organizational needs and cost factors, and thus these factors need to be a part of the theory foundation.

(2) *Solution Design*: In this activity, the fundamental need for technology creation or invention is thought out in detail. The development of solution technology is proposed with the help of notations for sequential steps and relevant flow diagrams. Software design and testing are done further in order to conform to the requirements. Two approaches for solution technology development include: re-engineering of an existing technology solution,

FIGURE 2.7:   Components of theory-building process.

and invention of a new technology solution with additional features. The process of development varies with the application area and the intent of the researchers who are involved in this activity.

(3) *Action Research Evaluation*: Once developed, the solution-specific technology is only assumed to be a means of addressing the problem until it is tested for usefulness. This testing of a technology solution takes up the evaluation of both the base theory and the technology. The evaluation is done on a map of three major areas:

a. Effectiveness and productiveness in eliminating a problem
b. Performance in comparison to other technologies for the solution
c. Compensatory impact on alleviating undesirable circumstances

Action research evaluation can be conducted using methods like field surveys, ethnography, and case studies. This form of evaluation is predominantly focused on problem solving for organizational needs. This may include identifying and choosing relevant but existing solution technology and its usage.

(4) *Simulation*: This is a form of evaluation in DSR process aimed at evaluating a solution-based technology in a controlled artificial environment. The methods adopted for non-real evaluation include laboratory experiments, field testing, and computer simulation intended to investigate the usefulness of technology in context.

These four activities are instrumental in shaping a framework for the DSR process. Offering a new dimension to the scientific design knowledge, DSR is regarded to be carried out in three broadly classified stages: Problem Identification, Artifact Design, and Evaluation. The following section offers details about each stage of the process.

2.7.1.1 Problem Formulation

In this initial stage of the process, a problem in question is identified. The source of the problem to be addressed may be the current challenge facing the business or how to monetize the possible opportunities created by new technology. This is triggered to accelerate business process operational efficiency and to find a way to accommodate new capabilities. In this whole process, a conceptual image of the problem and research objectives is sketched. The adoption of a particular method for problem identification [13] is decided on the basis of domain ontology, the study of objectives, problem-solving methods, requirement elicitation, and feasibility study. The final activity of the problem identification phase builds a research question and examines its relevance for practical problem solving. In brief, the process of problem identification can be further dismantled into the following activities: Identifying a Problem; Literature review; Expert interactions; Pre-inspection relevance.

a. Identifying a Problem
   DSR deals with practical problem solving. Thus, a research problem needs to be explicated to ensure its understanding and relevance to the problem domain. Conducting an empirical study enables researchers to find possible solutions to a problem. The problem should be generalized to make it

relevant and a common object of interest for more than one entity in different industry sectors wherever applicable.

b. Literature Review

For a generic problem identification strategy, a literature review is one of the most prevalent methods that can be used for research in multiple domains. Conducting literature reviews on scientific publications has the advantage of discovering unsolved problems, tapping new opportunities, and assuring relevance by generalizing the result descriptions. It is imperative to carry out literature studies in order to review its current state and possible impediments for its solution. A deep and informed literature research into practitioners' reports offers additional insights into suitable artifact building.

c. Expert Interactions

Expert interactions include one-to-one meetings with recognized domain experts and industry practitioners in order to identify domain-oriented unsolved and solved problems. Interviews and workshops are common ways of meeting with experts in the research problem domain. The expert survey may include the question: "Do you think the selected artifact offers a viable solution to the research problem?"

d. Pre-inspection

Once a relevant problem is identified in the research space, a pre-assessment on its relevance is conducted. This includes developing a research theory or hypothesis, suggesting a relationship between solution design and problem space. The theory foundation has been detailed in earlier sections of this chapter. A hypothesis or utility theory is evaluated in the following form:

- If a solution is applied, then this will improve the current state-of-the-art solution
- If a solution is applied, then this will reduce the problem space

The hypothesis has to be evaluated and adjusted continuously to represent the outcome of the entire research. The pre-evaluation on the relevance of the theory under consideration is done by taking the opinion of research practitioners from different disciplinary backgrounds, whether or not they are convinced of the hypotheses and presumptions supporting its applications in the real world.

2.7.1.2 Artifact Design

After identifying a problem and doing a pre-evaluation relevance check, a solution needs to be designed in the form of an artifact. This phase deals with rigorous research to postulate artifact design after analyzing the design cycle requirements. Literature research for design methods and system environment is carried out before selecting the artifact designing method.

During artifact design, the current state of the art of the existing solution needs to be taken into account. If the problem is restated, then the activities of the problem identification phase are iterated. For a clear understanding of the artifact design in context, design decisions should be reported in the most suitable format among all the entities. Methods for artifact design in scientific research can be traced as a result of situation analysis, task orientation, and synthesis. This phase of the DSR process

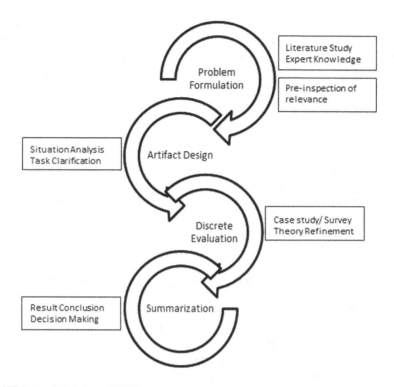

FIGURE 2.8: Workflow of DSR.

also involves a literature review of scientific publications to keep track of current trends in domain research as well as to respond to the changing needs of the industrial phenomenon (Figure 2.8).

2.7.1.3 Discrete Evaluation

This phase of DSR involves demonstrable observation of artifacts in context and validated integration of selected artifacts into the system design. Based on the evaluation criteria, the decision to adopt a solution to the problem is made. This activity makes use of hypothesis, theory building, method enhancement, simulations, and case studies to demonstrate the evaluation results.

Evaluation has to be reached to a sufficient state by means of naturalistic assessment or action research by conducting trans-disciplinary surveys, including experts showing general interest in the problem, and by controlled experiments in laboratories or soft simulations. A strategic approach to DSR evaluation consists of hypothesis refinement, case study, and research.

2.8 FRAMEWORK OF DSR EVALUATION

The fundamental construction of the evaluation framework in design science for research deals with the purpose of evaluation in functional aspects and also utilizes the paradigm of evaluation research. Built using the foundation of the purpose and the paradigm, this framework serves as a guide to researchers in planning a strategy for assessment of artifacts as well as utility theories they build during a research project. This also enables them to have a thought of further steps incorporating evaluation metrics to achieve DSR goals. An efficient evaluation strategy for a DSR project ponders upon why and when to commence evaluation after taking stock of problem components and of problems with evaluation.

2.8.1 DIMENSIONS OF EVALUATION IN DSR

Evaluation in DSR requires researchers to use well-executed methods of evaluation to present quality, efficacy, and performance of a design artifact in the context of change configuration. The process of designing an informative framework for DSR evaluation [22] is accomplished in the following phases: clarifying the objectives of evaluation process; selecting the strategies for evaluation; determining the characteristics of artifacts and hypotheses to evaluate; and creating segments of evaluation process components. A two-dimensional design of evaluation framework in DSR attempts to contemplate the validity of design theory and design artifacts specified for a problem domain.

Dimension X – Formative Evaluation

The formative evaluation in a DSR project deals with the functional goal of the evaluation and helps in refining the results of the task under evaluation. This is achieved by comparing deviation in produced outcomes: actual vs. expected, trailing the effectiveness of the process under evaluation. For an evaluation strategy to yield informed evidence, validation of knowledge outcomes by shared meanings and validation of results in the context of design theory should be judged.

Dimension Y – Paradigmatic Evaluation

This paradigmatic dimension [10] of the framework incorporates distinguishing between methods of evaluation adopted for a given situation in the research problem domain. In general, the transition of states in an evaluation process starts with the initial status of no evaluation, progressing toward understanding artifact evaluation in a realistic manner. Thus, doing chronological advancement through different modes of evaluation should represent the purpose fulfillment of the DSR project. The prominent modes of evaluation are artificial and naturalistic, by definition. While the increasing use of the artificial mode demonstrates a shared purpose with the common instinct of research, the naturalistic mode of evaluation tends to improve the knowledge quotient in progression and the effectiveness of artifact's utility by actual users. A planned projection of evaluations deemed valuable for the situation of the undertaken DSR project is known as the evaluation strategy. The following section walks through the different evaluation strategies and their selection for research in a particular design science project.

2.8.2 EVALUATION STRATEGIES

As a part of the strategy for any research project in design science, the pathway follows the intent of when, why, and how to evaluate a process in such a way that it should meet the defined requirements and shouldn't exploit the available resources. The formal strategies of a DSR evaluation [18, 22] capture the evaluation quadrants of rapid and comprehensive design, human likelihood and effectiveness, technical risks and efficiency, and core technical.

- *Rapid and Comprehensive Design*: This strategy lays focus on the relatively least objective evaluation and considers progression toward subjective evaluation relating to field studies, experiments, and artificial simulations. The trajectory of this evaluation strategy makes it cost-efficient and leads to a rapid conclusion of the project, but this may incur possible design risks.
- *Human Likelihood and Effectiveness*: This evaluation strategy focuses on objective evaluation or formative evaluation in early stages of DSR and progresses quickly to subjective evaluation near the end of the process. Subjective or experimental evaluation is aimed at examining the effectiveness of the design artifact in the real-time operational environment over the long run, despite the risks posed by human and social factors.
- *Technical Risks and Efficiency*: This evaluation strategy lays focus on the iterative artificial aspect of the design process and artifact, with progression toward subjective evaluation on the basis of simulation experiments. Unlike earlier strategies, the technical risks evaluation is done in iterations to determine the efficiency of an artifact in terms of mutual exclusion to other components of the framework. The success of this evaluation strategy implies that the usefulness of artifact is determined by the integration of artifact for a given situation and its sole impact, not of other factors. Nearing the final stage of this strategy, naturalistic evaluation factors are deployed.
- *Core Technical*: This strategy is implemented when the efficiency and efficacy of a design artifact are judged on a purely technical basis, not including human physiological or social factors. Essentially, this is used when artifact utility for product development is determined without the involvement of human users. The process is similar to a rapid and comprehensive design strategy except for the fact that this strategy necessitates only artificial evaluations for core technical artifacts throughout the process.

2.8.3 REAL-TIME EVALUATION SCENARIOS

a. Connected Data Ecosystem

The framework described above is comprised of archetypical or classical ways to achieve research goals. However, it is possible that cross-domain goal orientation may require more than one effective strategy or formulation of a hybrid evaluation. For example, in the world of the Internet of Things and connected ecosystems, one of the most pivotal areas of evaluation is clinical patient safety and real-time diagnostic reporting. A new kind of solution technology is needed to accommodate real-time communication for emergency cases between two physicians from anywhere at any time. This solution technology will meet the need for a connected healthcare ecosystem so that interaction between the clinician and the patient's external physician may happen in order to provide urgent care.

In this scenario, the design artifact should be created such that the goals of its technical utility and its socio-behavioral benefits are justified. Here, a hybrid evaluation strategy seems to be the best-fit approach in order to meet the design and technical requirements of a connected ecosystem. The possible ways of adopted hybrid evaluation strategy to work might include starting with a core technical evaluation (without human interference) to acquire knowledge about a new design artifact. Once the artifact for the clinical ecosystem proves its operational efficiency, the evaluation strategy may move forward with determining the efficacy of the designed artifact in overcoming human barriers and social risks. Imagining the situation illustrated in this example calls for the development of a project-specific evaluation strategy in the realm of DSR.

b.   Constraint-Based Project

Knowledge of the design challenge and its resolution is acquired from the creation and usage of the artifact, which is ultimately the end goal of the project in DSR. There is a wide scope of literature internal and external to the DSR field that pinpoints different methods for how to evaluate. In the different sections of this chapter so far, you have learned about the basic design process, DSR methodology, and the significance of an effective evaluation framework to reduce the impeding factors in a problem space.

The following section extends the framework of DSR evaluation [22] by including a four-phase strategy that can be developed for the synthesis of DSR outcomes in a constraint-based project, more commonly across multiple domains.

Step 1 – Be clear and specific with the goals:

There are certain engaging goals in building the evaluation factor of DSR. These goals are relevant yet varied at different stages of solution designing in DSR. Elucidation of goals depends on the determination of factors like rigor, design uncertainty and risk analysis, socioeconomic ethics, and efficiency.

Rigor: In a DSR project, rigor represents the development of artifact instance that catalyzes the process and stands mutually exclusive to external circumstances. Also the derived instance of artifact should prove its effectiveness by exhibiting the desired behavior in real-time situations. To recognize rigor in a DSR project, the artificial evaluation strategy is instrumental to evaluate the efficacy of an artifact, whereas expert-assisted evaluation proves best in validating benefits of the artifact in terms of its efficiency.

Design Uncertainty: The goal of functional evaluation becomes more significant when there is a possibility of design uncertainties impacting artifact performance in the DSR project. Thus, reducing the consequent risks by means of evaluation strategy becomes the foremost requirement. As discussed earlier, the kinds of risks in a DSR evaluation are identified as human risks, socio-behavioral risks, and organizational challenges that an artifact may fail to fit well in the use case of the project capabilities. Improving the design prototype of artifact supports the development of an effective and high-performance artifact, and also attempts to resolve the possible risks and uncertainties at the early stages of the process.

Socioeconomic Ethics: Designing a safety-critical solution technology should undergo evaluation to address the subsequent risks to the environment, people, organization, and project stakeholders. In addition, evaluation at the commencement of a project and at the final stage of its lifecycle is regarded as the best way to ensure that both artifact design and technology showcase par efficiency in dealing with all sorts of risks. Also, a combination of artificial and formative evaluation tends to sort the challenges facing teams in synthesizing knowledge at different stages of the research.

Efficiency: The goal of designing an efficient evaluation strategy is to balance all the above evaluation goals against the assets available for operational and functional evaluation. Specific techniques of evaluation take less time and complement the non-empirical evaluation methods to save time and cost efforts on research projects.

Step 2 – Choose the evaluation method(s):

Based on the objectives statement and artifact research, one or more evaluation methods may look more appropriate to solve a complex problem. Each evaluation method chosen for the solution should support the decision about how, when, and why to evaluate. For choosing an evaluation method [18] in a DSR project, the following heuristics or standard procedures can be followed:

Prioritize design risks: For a socio-technical artifact designed to address major uncertainties about social and behavioral challenges, and to develop long-term effectiveness in the real world, the human likelihood and effectiveness strategy would be most suitable. Likewise, if the major risk is about the likelihood of a technology to work in a certain manner, technical risks and efficiency method will help explicate the boundaries. Therefore, it is important to inspect and implement evaluation strategies in order to reduce potential risks.

Analyze the cost to organization: Individuals engaged in a design research project may interact with real stakeholders and analyze production systems to estimate relative costs by using a human likelihood and risk evaluation strategy. On the other hand, if it looks too costly for an organization to include real users and systems in the evaluation procedures, then a technical risk and efficacy evaluation strategy is pursued.

Analyze Time to Market: Evaluation based on the factors relating to artifact building for solution design also plays an important role in the choice of an evaluation strategy. If the artifact development is purely technical and addressing a problem needs deployment of the artifact in the far future, then a core technical evaluation strategy will work best. This puts forward the rationale of the negligible existence of the human settings in artifact utility with immediate effect. Similarly, if the design construct is small and simple without other risks incurred as above, then a rapid and comprehensive strategy can be approached.

Step 3 – Determine the attributes of artifact for evaluation:

The next step in choosing an evaluation strategy deals with the analysis of a basic set of features, objectives, and needs of artifact design. A detailed formulation of artifact properties and its purpose to tackle a situation is paramount during evaluation. Each artifact used in a given situation will rationalize the practical requirements and will invoke the common design theories in contemplation.

Step 4 – Perform strategic evaluation:
Having chosen a relevant strategy and worked out the artifact properties to evaluate, the real system evaluation series needs to be created. The rules for creating evaluation episodes for a given situation are defined based on the below workflow.

- Determine the environmental constraints and resource availability
- Prioritize the features to be included in context to the situation and problem
- Determine the number of iterations of evaluation and specific methods depending on the situation

2.9   DSR CYCLES

Understanding and evaluating the DSR process is incumbent not only to recognize its acceptance by professionals working on multidisciplinary projects but also to manifest the reliability of design science in research-steered domains in various fields. The DSR framework in Information Systems proposed by Alan Hevner [24] represents three cycles: Relevance, Rigor, and Design. This framework has been accepted by the majority of practitioners and design researchers worldwide and has been used as a benchmark in formulating objectives and processes of DSR [20] in a particular discipline.

a) *Relevance Cycle*: DSR is aimed at improving the system environment by introducing new tools, artifacts, and design practices. The Relevance cycle in DSR connects the activities of design science to the contextual elements of a research project. The relevance cycle in DSR takes account of the application's environment, not only identifying the research needs and complexity of a problem but also defining validation criteria for the acceptance of the research outcomes. Determining the frequency of iterations, the relevance cycle can be started in a research project by taking feedback from actual evaluation and new research requirement documents as inputs. As the first cycle of DSR, the relevance cycle deals with requirement elicitation and naturalistic testing to validate the need for change configurations.
   i. The requirement elicitation phase takes advantage of interactions between end users, domain experts, organizational components, and technical systems to collect business needs driving the research goal for a problem identified.
   ii. The naturalistic testing is governed by field analysis activities to test various scenarios pertaining to a problem space and to track new opportunities that lead to improvements in the system design practices. The field analysis of artifacts is performed to determine whether the design artifact is capable of improving the system environment, as well as whether or not the new artifact design lacks functional efficacy that may limit the use of the artifact in practice.

b) *Rigor Cycle*: This cycle establishes a relationship between theory building, scientific methods, and the design science process to address the objectives of a research project. Rigorous research is based on the shape and size of knowledge acquired out of scientific theories and engineering methods. The two major supplemental sources of this knowledge base are comprised of (1) Current artifacts and (2) Existing expertise and experience, in order to define the avant-garde favoring application domain. The rigor cycle in DSR governs innovation by providing existential knowledge to the research project. It depends on the researchers to minutely draw probing and inference to the knowledge base in order to ensure that the design as an outcome offers a substantial contribution to the research and is not a regular design for application in routine organizational systems. It is the rigor of research in design science that demonstrates the skillful selection and application of the appropriate theories, design methods, and evaluation of an artifact.

c) *Design Cycle*: The Design Cycle is central to the project of DSR. This component of the framework embodies artifact building, its evaluation, and resulting feedback in a research project against requirements until a high-performance design is achieved. It is necessary to understand the internal dependencies of the design cycle and mutual exclusiveness to other cycles during the actual research in execution. Thus, it is important to keep a balance of efforts consumed in building and examining the originated design artifact. Both activities must be convincingly based in relevance and rigor. Having a strong grounded argument for the construction of the artifact, as discussed above, is insufficient if the subsequent evaluation is weak. Rigorous testing of artifacts in field and laboratories calls for manifold iterations of the design cycle in the project before the artifact is released for contribution to the DSR cycle (Figure 2.9).

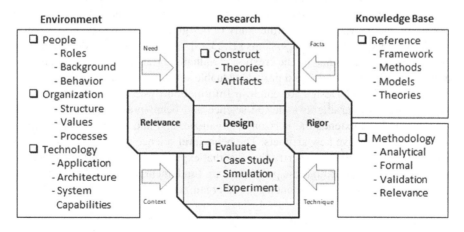

FIGURE 2.9: DSR cycles (adapted from [24]).

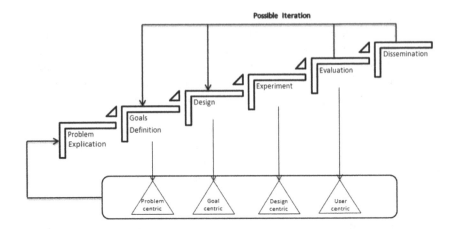

FIGURE 2.10:   Schematic diagram of DSR methodology.

2.10   DSR IN INFORMATION SYSTEMS: A CASE STUDY

It is generally accredited that design research involves the creation, inspection, and evaluation of artifacts in order to solve a real practical problem by using a scientific rationale. An artifact in DSR can be a construct, model, framework, method, or embodiment of a projected system. The study of methods, modus operandi, and best practices related to the design of an artifact and channeling its application for a problem space are incorporated in a type of research commonly known as design science. Therefore, the blending of the research question, problem explication, solution outline, creation and demonstration of artifact design, theory testing, and artifact evaluation can be collectively referred to as DSR, as shown in Figure 2.10.

The behavior and benefit of an artifact in a specific situation given in a problem space depend on the formulated research objectives and episodes of the design process. The fundamental principle of DSR can be applied to multiple disciplines having a common bias toward the scientific study of design artifacts and solution technology. In the field of information systems [21,23,24], the proposition of design science in research is underpinned by the creation of valuable IT artifacts that overcome the silos of insufficient theories and provide valuable solutions to domain problems.

The discipline of Design Research in Information Systems is characterized by two fundamental paradigms – design science and behavioral science. The design science paradigm extends the horizon of organizational and individual capabilities by forming innovative new artifacts. The behavioral science paradigm, on the other hand, attempts to build and verify theories that explain the organizational behavior and human tendency. Both these paradigms are fundamental to IS and are located at the convergence of people, technology, and organization.

Research in the information systems field examines more than just the technological system, or just the social system, or even the two side by side; in addition, it investigates the phenomenon that emerges when the two interact. – Lee A.S. (Editorial. MISQ 25, 2001)

Since the origin of the third industrial revolution, information technology has changed the way people consume resources to live, play, and entertain. The birth of the digital revolution in the era of the late 1950s to late 1970s marked the commencement of the Information Age. This revolution led the designers of computer science verticals to play a crucial role in ensuring that their research proves fruitful and their products provide value to web users. The DESRIST platform, for example, was created in 2006 to demonstrate the DSR happening in the information science research community.

Understanding the significance of DSR in the IS discipline [8,19], designers and researchers have come far beyond just collaboration to build various sorts of artifacts that not only catch the eyes of end users but also prove beneficial in providing the best experience of the application world. Many online platforms – learning, shopping, news, business, or whatever digital we browse today – have a great impact on our lifestyles. The increasing interactions between users over a plethora of digital platforms put multiple practical challenges before the IS design professionals.

ANALYSIS FRAMEWORK OF DSR IN INFORMATION SYSTEMS

Designing in information systems is viewed as a two-perspective process. In a remarkable study entitled "The Sciences of the Artificial", Herbert Simon advocated the concept of sorting the components in order to achieve satisfactory design goals. He stressed his belief that the construction of the design and the model of design process activities are important in design theory building. On the same side, Nigel Cross in "Developing a Discipline of Design Science Research" (2018) placed focus on design by doing, and gave more importance to the knowledge acquired from the designing process instead of theories.

Incorporating the above discussions of scientific research and design addressing the same research question, the following section of this chapter attempts to illustrate a referential analysis of DSR in information systems. This analysis presumes that people from different job roles in IT may commit errors in implementing the design theory and metadata in order to attain a common design objective. The advantage of using a metadata-based analysis framework lies in its ability to identify the patterns of differentiating results from individual studies and to furnish interesting combinations in the context of IS.

The contemporary analysis DSR model [14] in Figure 2.11 suggests that the study of problem space in a domain details about the nature of artifacts built in context,

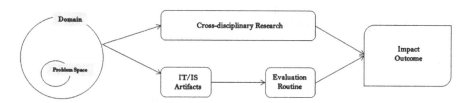

FIGURE 2.11:   Analysis of DSR in IS discipline.

the evaluation routines, cross-disciplinary research efforts, and its impact outcome. Each component of this analysis model is discussed further.

1. Domain and Problem Space
   The DSR for an IS project is carried out by identifying problem spaces from five primary research areas:
   a. Information Technology and Business
   b. Information System development
   c. Information Technology and People
   d. Information Technology and Market
   e. Information Technology and User Groups
   These areas are regarded as an apparatus of problem space comprising of generative (b) and declarative (a, c, d, e) symmetries for artifact classification, which resemble a system framework.
2. IT/IS Artifacts
   The characteristic output of the research cycle in DSR is referred to as an IS artifact. Different types of IT/IS artifacts are conceived of as construct method, model, and evaluation. Here the declarative symmetry is comprised of construct, method, model, and evaluation, and generative symmetry is comprised of refined design theories.
3. Cross-disciplinary Research
   Cross-disciplinary research is made functional based on the knowledge acquired from the study of existing benchmarks and research publications from outside and inside an institution.
4. Evaluation Routine
   The methods used to evaluate artifacts are categorized into the following groups: analytical, naturalistic, artificial, and socio-technical. The state of "no evaluation" in an IS research depicts the absence of the evaluation method to test an artifact.
5. Impact Outcome
   Impacts or relevance of a DSR outcome can be calculated in terms of contribution by utility and empirical citations. In the study of DSR or IS project, the usefulness of the framework is demonstrated by the comparative evaluation of research excerpts.

How to get down to selecting and building a relative strategy for different circumstances including social risks, along with differential and purely technical conditions, is a very important challenge in an IS project that needs to be taken up by design researchers and scientists.

SUMMARY

The constant state of change that characterizes the present time entails natural, socioeconomic, technological, and cultural aspects. Identified as a young cognition, design, in conjunction with other disciplines, deals with the aforementioned aspects as it evolves within the context of multidisciplinary practices. A design problem statement is the ingredient of innovative creation that brings science into action for the transformation of an existing design. This deep transformation involves not only the active areas of scientific research but also the different theories and knowledge that inform the choice of artifact in the design process. In the DSR, empirical and computational methods are primarily used to evaluate the efficacy and efficiency of the artifacts.

The study of people, design ontology, hypothesis, and research methodology pays the most important contribution to the exchange of design knowledge for practical problem-solving research in the real world. Accelerating the DSR knowledge synthesis outcome requires that research activities include theory building that is rigorously evaluated. A deep evaluation allows balancing the performance of the artifact in the design framework against cost and usability, which is critical to any design science project. The techniques for performance evaluation embrace problem identification, risk analysis, design methods, and acceptance testing. Different stages of evaluation establish the degree of artifact utility and the standard of knowledge offered by design science.

DSR contribution in Information Systems is attributed in its methodology to implement business needs in an appropriate environment and in the ways in which it adds to the content of the design knowledge base for further research and practice. Rigor and relevance in the DSR cycle strengthen the balance of structure and creativity that emerges from accumulated knowledge. Furthermore, a systematic view of research methodology in a design science project provides knowledge about the usefulness of an artifact to solve an existing problem.

REFERENCES

1. Guohai Zhang, Yusheng Li. "Multiple Disciplines Product Design Knowledge Representation Strategy Based on Ontology and Semantic Network" *Telkomnika Indonesian Journal of Electrical Engineering* 11(10) (2013): 6074–6079
2. Zech Andreas, Ramsaier Manuel, Stetter Ralf, Niedermeier Hans-Peter, Rudolph Stephan, Till Markus, Holder Kevin. "Model-Based Requirements Management in Gear Systems Design Based on Graph-Based Design Languages" *Applied Sciences* 7 (2017): 11
3. Lucienne T.M. Blessing, Amaresh Chakrabarti. *DRM, a Design Research Methodology*, Springer (2009)
4. Stephen Green, Darren Southee, John Boult. "Towards a Design Process Ontology" *The Design Journal*, Routledge 17(4) (2014): 515–537
5. Mario Štorga, Mogens Myrup Andreasen, Dorian Marjanović. "The Design Ontology: Foundation for the Design Knowledge Exchange and Management" *Journal of Engineering Design*, Taylor & Francis 21(4) (2010): 4
6. Richard Baskerville. "What Design Science Is Not" *European Journal of Information Systems* 17(5) (2008): 441–443
7. Robert Winter. "Design Science Research in Europe" *European Journal of Information Systems* 17(5) (2008): 470–475
8. Vijay Vaishnavi, Bill Kuechler. *Design Science Research in Information Systems*, DESRIST (2012)

9. Roel J. Wieringa. *Design Science Methodology for Information Systems and Software Engineering*, Springer (2014)

10. Juhani Iivari. "A Paradigmatic Analysis of Information Systems as a Design Science" *Scandinavian Journal of Information Systems* 19 (2007): 39–64

11. N. Cross. "Science and Design Methodology: A Review" *Research in Engineering Design* 5(2) (1993): 63–69

12. N. Cross. "Designerly Ways of Knowing: Design Discipline Versus Design Science" *Design Issues*, MIT Press, 17(3) (2001): 49–55

13. Olga Levina, Udo Bub, Marten Schonherr. "Outline of a Design Science Research Process" *Proceedings of the 4th International Conference on Design Science Research in Information Systems and Technology, DESRIST*, ACM Press (2009)

14. Olusola Samuel-Ojo, Doris Shimabukuro, Samir Chatterjee, Musangi Muthui, Tom Babineau, Pimpaka Prasertsilp, Shaimaa Ewais, Mark Young. "Meta-Analysis of Design Science Research within the IS Community: Trends, Patterns, and Outcomes" *Global Perspectives on Design Science Research*, Springer (2010): 124–138

15. Dresch Aline, Daniel Lacerda, Pacheco Antunes and José Antônio Valle. *Design Science Research, a Method for Science and Technology Advancement*, Springer (2015)

16. John Venable. "A Framework for Design Science Research Activities" *Proceedings of the Information Resource Management Association Conference*, Idea Group Publishing (2006)

17. Vladimir Hubka, W. Ernst Eder. *Design Science: Introduction to the Needs, Scope and Organization of Engineering Design Knowledge*, Springer (2012)

18. John Venable, Jan Pries-Heje, Richard Baskerville. "A Comprehensive Framework for Evaluation in Design Science Research" *Proceedings of the DESRIST 2012; Design Science Research in Information Systems. Advances in Theory and Practice*, Springer LNCS 7286 (2012): 423–438

19. Alta van der Merwe, Aurona Gerber, Hanlie Smuts. "Guidelines for Conducting Design Science Research in Information Systems" In: Tait, B.., Kroeze, J., and Gruner, S. (eds.) *ICT Education. SACLA 2019. Communications in Computer and Information Science*, vol 1136. Cham: Springer.

20. A.R. Hevner. "A Three Cycle View of Design Science Research" *Scandinavian Journal of Information Systems* 19(2) (2007): 87–92

21. Ken Peffers, Tuure Tuunanen, Charles E. Gengler, Matti Rossi, Wendy Hui, Ville Virtanen, Johanna Bragge. "The Design Science Research Process: A Model for Producing and Presenting Information Systems Research" *International Conference on Design Science in Information Systems and Technology*, DESRIST (2006): 83–106

22. John Venable, Jan Pries-Heje, Richard Baskerville. "FEDS: A Framework for Evaluation in Design Science Research" *European Journal of Information Systems*, Taylor & Francis (2016)

23. A.R. Hevner, S. Chatterjee. *Design Research in Information Systems: Theory and Practice*, Springer Publishing (2010)

24. Alan R. Hevner, Salvatore T. March, Jinsoo Park, Sudha Ram. "Design Science in Information Systems Research" *MIS Quarterly* 28(1) (2004): 75–105

25. Paul Johannesson, Erik Perjons. *An Introduction to Design Science*, Springer (2014)

26. Alan Hevner, Jan vom Brocke, and Alexander Maedche. "Roles of Digital Innovation in Design Science Research" *Business & Information Systems Engineering, Springer* 61 (2019): 3–8.

27. Case, Jenni, and Gregory Light. "Emerging Research Methodologies in Engineering Education Research." *Journal of Engineering Education* 100(1) (2011): 186–210.

3 Art of Science in Fashion Design and Technology

3.1 OVERVIEW

Any innovation in the world is an evidence of the forces that combine to produce significant results. Tapping possible developments in engineering, fine arts, information systems, and the fashion and textile sector, the study of science is the only pivotal force behind the discovery of new solutions. There is no unattended field in which science doesn't drive the gap analysis for sustainable development. As an example, in the apparel industry, science in design supports artists and fashion and textile designers in recognizing the needs of customers and deriving new solutions for known challenges.

Gaining new scientific knowledge throughout the design lifecycle helps designers and manufacturers to improve production methods like weaving, printing, dyeing, pattern making, branding, and labeling. Also, it helps businesses to provide dynamic people with the creative abilities necessary to up skill and support the sector. The following sections walk through the role of science and technology for fashion as a design process and product, by drawing on a few of the scientific and technological trends of current and futuristic fashion segments.

3.2 CONNECTING RELATIONSHIP BETWEEN SCIENCE AND FASHION

The descriptive foundation of science in fashion has immense potential to assert the significance of the two buzzwords as they go hand in hand interlaced with each other. The fusion between science and fashion motivates the research of designers and scientists to innovate new promising solutions that tend to meet the demand of customer experience and industry needs as well. The appropriate blend of science and fashion puts forward an excellent opportunity for designers and scientists to create a blueprint for their research objectives.

Using the principles of scientific knowledge to create a fashion masterpiece, designers can advance their skills and showcase their talent on a broader platform. The focal point for research by fashion designers starts with collecting the images of interest, exploring new ideas, and triggering innovative design solutions by taking inspiration from the source. In general, the major sources of inspiration for designers include a scientific body of knowledge and existing discoveries that help them design their final sketches.

3.3   WHAT MAKES A GOOD DESIGNER?

Design embodies natural resource extraction, personification, technologies, and social codes that delegate the innovative culture to prosper. It's often a topic of deliberation for people around the world: *what the traits of an intuitive design are*, and *what truly makes a good designer*. This calls for a search to find answers and seek out solutions to real problems that great designers solve.

The world becomes a better place when we understand the surrounding ecosystem around us with a wider lens and make innovation work for mankind. Finding different ways to solve a practical problem, a designer develops innovative thinking and makes note of things to build a better solution that others often overlook. This makes room for science to come into action and make technological innovations possible for the fashion design world.

3.4   FASHION DESIGN AND FASHION TECHNOLOGY

The origin of fashion starts when people adorn their look with dress and style. This happens in the hour of the morning when they dress themselves and prepare for the day. According to "Moriarty" – "Fashion is what you wear and how you want to present yourself to the world". However, a sense of satisfaction only touches your soul when you style with artistic yet comfortable wear. This aesthetic and comfort is the key to successful fashion design. The following section describes the expression of ideas, the fashion life cycle, communication, the retail marketplace, and how we can realize the basic essence of design science in fashion technology. These key stratagems, when recognized, allow designers tap the technological possibilities of improving garment production and market value.

Nowadays, fashion design and fashion technology go hand in hand, demonstrating their effectual outputs collectively. In order to achieve tangible assets in their avant-garde endeavors, the fashion industry is relying on the power of these two buzzwords to deliver exceptional and delightful customer experiences in stores and on online platforms with a large number of shopping marketplaces. Pushing deeply into design science in fashion technology requires understanding fashion terminology and the relationship between fashion, science, and design, after grasping the knowledge of fashion design and fashion technology.

Fashion design is the art of applying design and aesthetics to clothing and its accessories that gives someone an identity, which also mirrors cultural and social attitudes varying in different places and times. Technically, this term deals with the designing aspect of fashion and the apparel industry, in which fashion designers focus on curating innovative clothing sense and creating products. On the other hand, **fashion technology** is the technical parameter of the fashion domain, which is focused on the development of new technology for the production of innovative fabrics by using new methods of dyes and prints.

Fashion designing is the art of customizing a dress or other apparel by exploring the different relational possibilities between performance, culture, art, and environment. On the flip side, the objective of fashion technology is to increase the

TABLE 3.1

Comparison between Fashion Technology and Fashion Designing

	Fashion Designing	Fashion Technology
Definition	Fashion designing is the creative field in which trendy and appealing apparel or accessories are designed	Fashion technology covers the manufacturing process and involves a wide usage of technology in the production of apparel
Aspect	More creative	More technological
Specialization	Leather design Accessory design Textile design	Generally it provides specialization in apparel manufacturing and information technology
Spectrum	Analysis of consumer trends Designing clothes and accessories Knowledge of fabrics Ability to combine colors, shades, and textures Sketches of the design, etc.	Covers the dimension of sewn product manufacturing industry: Understanding of fabric and manufacturing process Apparel quality management Technology for production Fabric processing, production process, etc.

buying motivation of customers, and to catch up with trends or styles that have been designed by the designer in the domain of fashion designing. It covers the manufacturing process and involves a wide usage of technology in the production of apparel. Undoubtedly, fashion designing has become one of the most popular courses among students, as it combines the style and beauty of the fashion world. In this world, where many types of skills are required to work on designing cloths and accessories and to invent innovative styles, the design process is equally important. Knowing the process of the product or changing the raw material for the design would be beneficial in rectifying the design if it doesn't meet with approval.

The role of the fashion technologist comes into play after getting approval for the designs from the designer. A fashion technology expert provides the necessary instructions on the way these products will be designed by considering the type of fabric and its suitability for the final product. Therefore, the art of fashion design [6] is incomplete if we do not involve the fashion technologist, who primarily works on the core garment manufacturing technology and garment production. However, there is a boundary between these two fashion terms that spotlights their differences for better understanding of the nature and characteristics of the complete process (Table 3.1).

3.5 SCIENCE IN FASHION DESIGN

Scientific developments play the role of assessing whether the innovations actually improve standards. With rapid change in the society and market ecosystem, technological innovations are reshaping the world at a much faster pace than ever before.

The combination of design research and innovation culture has swept away many global companies in a tidal wave of progress. The rate at which technology is developing has fundamentally changed the outlook of the way any business operates. Nevertheless, it is because of scientific bodies that the goal of sustainable fashion, once being seen as nice to have, is now more tangible to achieve.

In the era of science and technology, global collaboration has led to a plethora of ideas that are contributing to lifting the world of fashion by three powerful forces: Science, Design, and Technology. Utilizing science as the foundation in design pedagogy has empowered businesses to put their design innovations on the global streets while meeting organizational needs during changing times. Scientific development guides technological innovations, which in turn address transformation challenges facing the fashion industry.

"Technology is indeed a queen: it does change the world" (Braudel, 1981)

Assuming the similarities between the worlds of Fashion and Science, scientists as well as designers could work in the same direction with a common goal to create something unprecedented by introducing change for innovation. This realization encourages many intellectuals from different fields to synergize their efforts and take advantage of tangible science to create innovative designs.

In general, people are of the opinion that science and fashion are two different tangents of design world and exist in opposition to each other. However, it is hard to believe that the absence of science is empowering, when it is evidently paving the path for designers to truly understand a design without consuming a lot of time in creating hundreds of sketches to reach the final selection. The majority of people may find that creative vision is the only necessity of good designing, but nonetheless it is likely that the scientific foundation behind what we know as symmetry makes a balance between visual aesthetics and uniform design parameters. Therefore, in developing a garment, a good designer will always leverage those opportunities which boil down fashion into systemized methods but do not pull the glamour out of it. Having this, seeing fashion through the lens of the scientific approach also transforms the journey of design industry in three ways: psychology, innovation, and reach [8].

1. **Psychology**
   By acquiring knowledge of fashion psychology, understanding the "hows" and "whys" of a design strategy, designers are able to justify what appeals to users and why. This psychological study of human mind mapping helps to improve product standards in all aspects of the industry.
2. **Innovation**
   Innovation is easily understood as the application of better solutions that meet new requirements, unarticulated needs, or existing market needs. In different contexts, an innovation can be a new idea, a creative thought, or a new imagination in the form of product or style. According to Pete Foley, the innovation process is *"a great idea, executed brilliantly, and*

communicated in a way that is both intuitive and fully celebrates the magic of the initial concept". Untouched by the field of fashion design, innovation has different processes, such as development of strategy, method, instrument, technique, and material, and is not concerned only with the development of new garments. Overall, innovation is an application of the scientific design framework that leads creativity to thrive in its full dimensions.

3. **Reach: Understanding why we wear what we wear can help skeptics engage with fashion**

To understand the scope of product design, reaching out to people surely helps in understanding the interests, needs, and trends of fashion wearers. Knowing what and why people wear certain styles of garments enables designers to actually engage their line of thinking with the fashion world. This also helps in improving the use of fashion psychology. This implies assessment of consumers' personality traits and exploration of aesthetics that fits best in respective manner.

Nonetheless, it's the collaboration of fashion and science that has set a benchmark by inventing many impeccable products to meet the demand of industry and customers.

3.6   ROLE OF TECHNOLOGY AND CREATIVITY IN FASHION DESIGN

The venture of technology and creativity in fashion design not only boosts the possibilities for fashion designers but also creates a space for technocrats to synthesize the world by utilizing technology in today's environment. When discussing the role of creativity in fashion design, it is the mantra of modern apparel industry and it becomes important to communicate if a designer's collection has been made with a focus on rich creative design. For beginners, it is very difficult to learn the skills of fashion designers like sketching, pattern making, and drawing, but due to technology newcomers need not to have inherent skills, as after learning the application of technology, they can also be excel in the same field. The new technologies make creativity international.

3.6.1   TECHNOLOGY

Recent studies establishing the relationship between technology and the fashion industry have led to new categories of technology, namely, Process Technology and Information Technology.

Process Technology: Process Technology relates to the hardware and software that enable the product to be produced. This category has been further subdivided into CAD (computer-aided design), CAPM (computer-aided pattern-making), CAM (computer-aided manufacturing), and 3D Body scanning/mass customization.

- *CAD*: It encompasses programs that facilitate creative sketching, presentation boards, technical design, and textile design such as Adobe Photoshop and Illustrator, In-Design, and U4IA.
- *CAPM*: It includes digitizing existing patterns for grading and pre-production preparation, creating patterns from existing slopers, digitizing draped muslins for completion on the computer, customizing patterns for made-to-measure clients, and the newest development, 3D virtual assembly of the pattern.
- *3D virtual environment*: This software enables evaluating the pattern on a virtual model to see the virtual model walk, allowing changing of color, pattern, proportion, and details of the design before realizing it in actual fabric.
- *CAM*: It includes cutting a large quantity of garments in a paperless environment, using automated spreading and cutting, and in some segments of the industry, going all the way to automated assembly. It also includes supply-chain technology such as software for Product Data Management and Product Lifecycle Management (PLM), both in a local network and Internet-based.

Information Technology: Information Technology is a cluster of computer technology and communication technology. Information Technology takes care of the hardware and software part, whereas Communication Technology takes care of the transmission part. The Convergence of information technology and communication technology has brought dramatic changes in routine lifestyles by converting the physical world into "Human-ware", a term that depicts the design of software and hardware aimed at improving human interactions with the system, also known in general as user experience. This new zone of human-ware connects people beyond geographical boundaries via virtual meeting apps, VOIP networks, collaboration tools, and open communities. Thus, information technology is not only a computer technology including hardware and software for processing and storing information but also a communication technology for transmitting the information from place to place, person to person, and organization to organization.

Over the decades, Information Technology has been utilized in every aspect of fashion designing to keep up pace with the changing trends of the market situations. Various forms of Information Technology, including but not limited to simulation of prints and fabrics, are applied in the principles of designing, fashion marketing, and color rendering to make and focus on multiple dimensions of fashion creativity through computers. This is largely supported by the advancements Internet services that enable communication between creators of fashion designs and manufacturing companies across the world. With the emerging and ever-increasing use of software tools to innovate in design processes and design technology, design professionals and manufacturers are now able to complete their jobs in time with more responsive means of delivering products, ultimately reducing time-to-market and production costs while increasing their return on investment. Figure 3.1 demonstrates clusters of both technologies, meeting at a common focal point: the Internet.

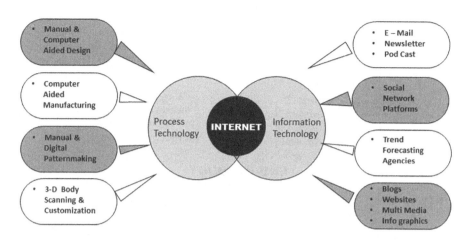

FIGURE 3.1:   Typology of technology in fashion design.

3.6.2 CREATIVITY

Creativity in Fashion Design applies to all companies or individuals involved in the creation, production, and merchandising of items distinguished by their aesthetic and functional properties. This in turn triggers learning about psychological instincts related to the desires of people consuming the items. The wave of digital transformation in the fashion industry supports the fact that the sutras of creativity in fashion design are inextricably linked to user-centric approaches for design.

The type of creative technique employed in fashion designing depends on the problem to be solved, which in most cases directly corresponds to requirements of the customers interested in buying the product. When examining creativity in fashion design, two distinct categories of creativity emerged: Leadership Creativity and Adaptive Creativity.

Leadership Creativity involves focusing on highly refined techniques, materials, and craft to develop a product that is fashion-forward, innovative, and directional.

Adaptive Creativity involves developing a product with well-established parameters or limitations, especially those related to price and production, exacting a heightened awareness of operations, management, methods, and technology.

Leadership Creativity and Adaptive Creativity have eight comparable components of design. These design components, which link to the metrics of product evaluation, are as follows: (1) Research Initiatives and Investment, (2) Selling price, (3) Product Attributes, (4) level of consumer taste, (5) techniques for creation, (7) consumer perception of product lifecycle, and (8) source of design inspiration. Figure 3.2 provides a visual depiction of the typology of Leadership Creativity and Adaptive Creativity. On the basis of the design components, each kind of creativity generates a viable product for the appropriate market.

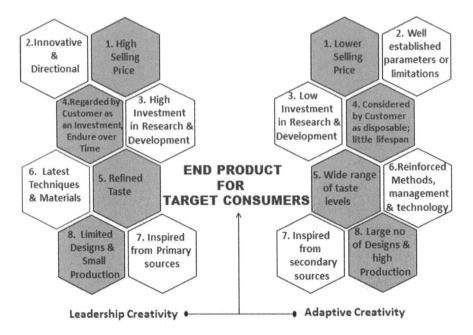

FIGURE 3.2:   Typology of creativity in fashion design.

Careful examination of the eight components of this typology shows that each component has distinct characteristics according to creativity type.

Product Design: Use of Leadership Creativity demands significant investment in research and development, and designers need to look to abstract sources for inspiration. On the flip side, using Adaptive Creativity lays less emphasis on research and development with less investment, and designers can find inspiration from secondary sources, most often by investigating already manufactured garments.

Market Value: A Product with Leadership Creativity (PLC) is assumed to be sold at a high price, for the ground fact that the product is produced in low numbers to maintain exclusivity, requiring designers not to create more samples of designs per season. A Product with Adaptive Creativity (PAC) is expected to be sold at a mid-range to low price with considerable emphasis given to establishing efficient management of operations, materials, and production methods while maximizing the potential of Process Technology to bring costs down. The designer of the PAC is expected to generate a high number of designs per season that meet strict requirements of price and consumer demand.

Taste of Consumer and Production: PLC is produced in limited quantities, appealing to a customer with refined taste who considers the purchase as an investment. PAC is produced in large numbers, appealing to customers with a varying taste levels ranging from refined to popular, who consider products that have short lifespans and are easily disposable.

3.7 SCIENCE AND TECHNOLOGY IN FASHION DESIGN INNOVATION

Let's learn how the most interesting technology bridges between science and fashion have influenced the trends of garment design.

1. 3D PRINTING TECHNOLOGY

3D printing technology is the construction of a three-dimensional object from a CAD model or a digital file. The term "3D printing" can be referred to as additive manufacturing, in which an object is created by laying down successive layers of material such as liquid molecules or powder grains being fused together. There are various areas of different industries like fashion, aerospace, automotive, food, gaming, education, aviation, dental, and health care where the 3D printing technology [10] has been continuously flourishing and likely to reach a market of $41 billion by 2026 (Figure 3.3).

In the fashion industry, designers are really pushing their newest ideas into reality by utilizing and synthesizing advanced materials and performing rapid prototyping of 3D products. However, 3D-printed fashion products are not yet largely visible at local fashion stores, but they have marked their unprecedented rank in driving innovation to runways. Initially, the scope of this technology was limited to designing the jewelry, and sole of shoes and sporting footwear due to the non-flexible nature of the materials. With time and extensive research and development, it has become possible for designers to develop different forms of 3D garments, for example a jacket designed using a 3D printed mesh system which can be later assembled as desired.

2. STITCHLESS TECHNOLOGY

A garment that is made without using stitches is known as a stitchless garment. The parts of such a garment are joined together without using a sewing machine or stitches. Instead, these garment parts are fused together by means of heat sealing and bonding technology. One of its modern developments – seamless garment technology – allows designers to use specialized circular looms to develop adaptive garments that fit the body size (Figure 3.4).

FIGURE 3.3: Products made by 3D printing technique.

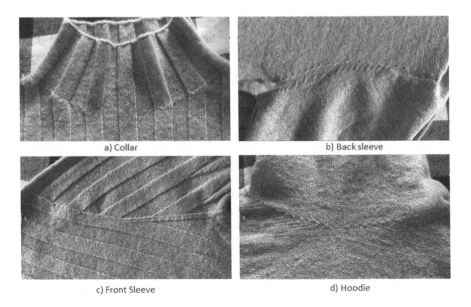

a) Collar   b) Back sleeve

c) Front Sleeve   d) Hoodie

FIGURE 3.4:   Examples of seamless cardigans.

Seamless garment technology [12] brings benefits to the business in terms of higher production rates, quicker response to market demands, reduced cost and time, and elimination of worker-oriented process for fabric laying, cutting, and sewing. Stitchless or seamless garment manufacturing technology finds its applications in various verticals of the textile and fashion industry, including but not limited to the following:

- *Protective textiles*: Garments that are designed to protect from external damage or harsh environmental conditions are nowadays produced using seamless filament-knitting techniques. Using this technique, apparel and wearables like gloves are made lightweight and flexible to be consumed in the electronics, food-handling, paint, plastics, and other high-precision business sectors where safety measures are paramount to ensure wellbeing of the workers in addition to hygiene and cleanliness.
- *Ready-to-wear Garments*: Seamless apparel production technology offers both the aesthetics and the functional components needed to create new products for customers. The resulting garments provide comfort during those times when the wearer wants to relax. Seamless technology can be applied to active wear, leisurewear, underwear, and sleepwear products.
- *Second-skin Textiles*: Seamless apparel construction is aimed at supporting free muscular movement. The novel development of advanced sports-oriented textiles is based on the seamless garment technology to provide an easy fit by use of elastic and quick-drying yarns. Like active wear and

swimwear, a diverse range of products such as gloves, hats, and socks are some of the more obvious applications in this category.

- *Cardio-sensing Textiles*: A new kind of sportswear available in the market – the seamless sports bra – exhibits heart-sensing smart technology. Seamless cardio-sensor placement is given to offer greater support during high-impact sports. The underlying textile electrodes are knitted into the bra and placed such that they can stretch and capture the electrical signal of heart's pulse during movement by the wearer, while maintaining contact with the skin.
- *Medical Textiles*: In medicine and the life sciences, seamless designing technology has offered numerous advantages by incorporating high-performance fibers and additional sensors with a blend of specialty fibers. Seamless textiles provide the required healing functions of the garment as the situation demands. Examples are bandages, orthopedic supports, and medical compression stockings.

3. PARTICLE TECHNOLOGY

Particle Technology is one of the most innovative and latest revolutions in the world of fashion design. Construction through particle technology uses instant spray-on materials made from polymers, and natural and man-made fibers. The latest spray fabrics are prepared by spraying on the body using aerosol technology. The spray is delivered from a compressed air spray gun or aerosol can and dries instantly on contact with the air, creating a non-woven fabric for many surface applications.

The spray solution [5] contains small fibers blended with polymers that hold all fibers together and a solvent that releases the fabric in liquid and disappears as soon as the spray touches the surface. The category of fabric that is produced is based on the non-woven technique because the proper thickness of non-woven fabrics can be achieved as per the requirement of the garment. The texture of the fabric can be altered by adding or changing the blend of the fibers (linen, acrylic, and wool) that are used in the spray. To increase the thickness volume of spray fabric, multiple layers of spray are used across the required region of the garment. The technology is expected to find creative applications in the fashion industry (Figure 3.5).

This spray has also potential applications in the medical field in applying dressings on burnt skin without exerting much pressure. In the automotive industry, the spray can be used where light, stain-resistant, and durable fabrics are required. These fabrics enable creative shapes and biometric forms, although they feel very cold when sprayed.

4. 3D MODELING

3D technology has been used to envision designs. 3D modeling leverages the development of a mathematical illustration of the three-dimensional surface of objects via specialized software. 3D modeling software offers the creation of new designs while testing prototypes in parallel, which makes it easier to define simulations of space

FIGURE 3.5: Fabrication through aerosol technology.

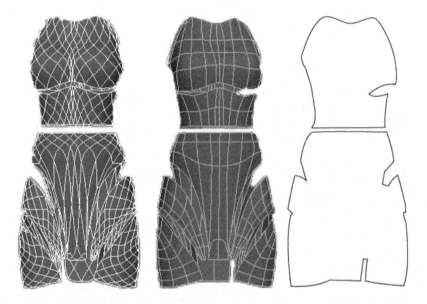

FIGURE 3.6: Demonstration of 3D modeling technique.

and fashion design. The launch of simulation software by brands like LO3D, Tuka tech, and Optitex has enabled artisans and fashion designers to create dynamic 3D garments on the computer (Figure 3.6).

5.   BODY SCANNING

When science, technology, and fashion come together, this can dodge many great hurdles in shaping new garments with unique features. The superlative applications of three-dimensional technologies available in the market – for example, body scanning – combine the properties of specific fabrics, body shapes, and stitching patterns to determine the exact measurements of garments.

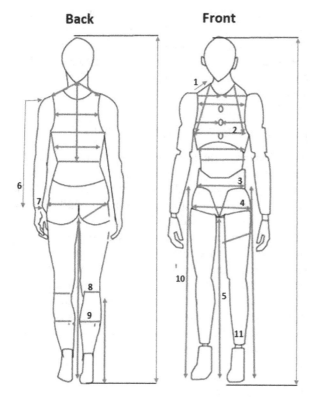

Back   **Front**

1. Shoulder
2. Chest
3. Waist
4. Seat
5. In-seam
6. Sleeve Length
7. Wrist
8. Knee
9. Calf
10. Length W-B
11. Bottom

FIGURE 3.7: Measurement through 3D body scanning.

In the 3D body scanning technique, wearers are scanned to know their correct size and thus make it easy for brands to determine the correct size, saving them time and money (Figure 3.7).

The 3D scanner is equipped with a white light that captures the most accurate measurement of the body. This information is collected and processed to accurately match the measurement of the size and tightness of the garment. The obtained fitting is inspected using pattern-making software, and final changes in the measurements are made on the paper patterns. In this way, multiple aspects of body dimensions are calculated for quantitative study of garment size and improving the fit measures of body figures. Modern 3D scanning applications include mechanical functions designed to reduce the cycles of iterations and allow expert technical designers to alter patterns quickly [8].

6. SMART TAILORING

Conceptualized by Siddhartha Upadhyaya, an Indian designer, smart tailoring has been launched as a new process of innovation in the fashion industry. Also known as direct panel on loom technology (DPLT), this concept gives fashion designers aid in making intricate and continuous designs for complex garments with less consumption of energy, fabrics, time, and money. This technology attaches the looms directly to a computer in a way that any information regarding color, pattern, or size of a garment is interpreted by the loom and it then calculates the precision to the number of required pieces. This eventually saves a lot of time in queuing the samples, weaving, fabric cutting, production planning, etc.

3.8 ARTIFICIAL INTELLIGENCE IN FASHION DESIGN

In the recent technological developments, the use of artificial intelligence has been an integral part of the transformational change in the world of fashion. As discussed in the above sections, fashion design is an art, and most "classic" fashion designers still work with sketches and mood boards to craft a user-inspired, market-friendly design. However, with the advancement in science of fashion and the use of the latest technology in the design process, the tools used by fashion designers are also becoming digital. Thus, an emerging generation of so-called "hybrid fashion designers" is hailing; these designers and creators develop digital forms, conceptualized with the help of social media, industry trends, and Artificial Intelligence enabled technologies. The following section is dedicated to studying the relevance of technologies within the fashion industry, and the ways in which top fashion brands are leveraging technology-driven innovations like artificial intelligence and robotics to improve their production technology and market outreach.

3.8.1 PRODUCTION TECHNOLOGY PERSPECTIVES

- *Data Collection*:
    With more sophisticated data collection, fashion brands are using technology to understand customer needs and design better apparel. For example, a Germany-based fashion platform, Zalando (in partnership with Google), is using AI-powered fashion designing that is based on the customer's preferred colors, textures, and other style preferences.
- *Apparel Design*:
    The best product design work starts with a creative sketch in an apparel designing. Traditionally, when a design was not approved, the team had to go back to the drawing board to create a new board, which took a lot of time and money. But the emergence of creative software like Adobe Suite, SmartDraw, Wacom, and Techpacker [2] has allowed designers to outperform manual drawing techniques. This has attracted numerous technology-savvy designers to use digital design toolkits that offer excellent smart layout extensions and allow designers to create clothes with infinite flexibility on virtual models to meet their goals more efficiently (Figure 3.8).

Once the sketch has been approved, then AI provides various tools to reimagine or rework the look of designs by draping virtually varied patterns and fabrics available in their toolbox. And all this is done in a matter of minutes, if not seconds. This in

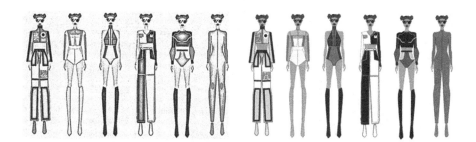

FIGURE 3.8: Manual and digital sketching.

turn also speeds up the supply chain process by getting products to the shelves in weeks to keep up with the consumer demands. On the other hand, some companies use "genetic algorithms" to design new apparel styles with exhaustive details of color, sleeve style, and hem length on their websites or to create new styles by recombining attributes from existing styles and possibly mutating them slightly for their own brands. The other application of AI which it helps people is to define their personal style by recommending outfits from their own closets. Though many of the AI solutions being explored in the fashion world are still in their infancy, they're only going to improve with time. It's a disruptive technology likely to impact even the stereotypically set-in-its-ways decorated-apparel business.

3.8.1.1 Technology for Fashion Designers

- *Development of Techpack*

  The major frustration of a fashion technologist comes when he works on correcting flat sketch measurement errors, redrawing the specification sheet for repeated data entry to fit in a format. This consumes a lot of hours and makes experts tired of exerting effort in monotonous practices. There are a lot of softwares available in the market that can put the details into one sheet with all necessary sketches and notes. Among these, Techpacker is one of software that fulfills the needs of the hour. It uses an advanced card technology that dominates fashion PLM (according to Mark Harrop, CEO and MD of WhichPLM).

  Developing a techpack without any error through technologies not only saves time for merchandisers but also helps in building an image of any organization by finishing their target on time.

  Techpacker breaks down techpacks into cards that can be dragged and moved to any desired location. A user can copy, move, connect, and reuse the components countless times before reaching the final design (Figure 3.9).

- *Manufacturing Process*

  There are certain problems often faced by companies in order to deliver their products on time to market at low cost; these problems include but are not limited to conducting deep market research, producing high-quality

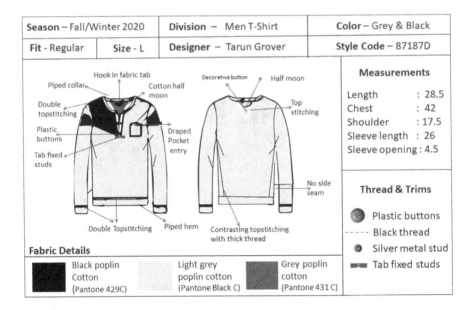

Season – Fall/Winter 2020		Division – Men T-Shirt	Color – Grey & Black
Fit - Regular	Size - L	Designer – Tarun Grover	Style Code – 87187D

Measurements

Length   : 28.5
Chest   : 42
Shoulder   : 17.5
Sleeve length   : 26
Sleeve opening : 4.5

Thread & Trims

Plastic buttons
Black thread
Silver metal stud
Tab fixed studs

Labels on diagram: Hook in fabric tab, Piped collar, Cotton half moon, Decorative button, Half moon, Double topstitching, Top stitching, Plastic buttons, Draped Pocket entry, Tab fixed studs, No side seam, Double Topstitching, Piped hem, Contrasting topstitching with thick thread

Fabric Details

Black poplin Cotton (Pantone 429C)	Light grey poplin cotton (Pantone Black C)	Grey poplin cotton (Pantone 431 C)

FIGURE 3.9:   Techpack for garment specification [12].

products, complying with industry regulations, overcoming bottlenecks on the floor, and iterative testing. To mitigate these problems, some leading manufacturers in the industry have come forward to make use of AI and machine learning tools that allow them streamline manufacturing and business operations in response to fast-changing fashion trends. To name a few, leading fashion brands like Zara, Tommy Hilfiger, and H&M are reaping the benefits of improved efficiency and decreased downtime by recognizing seasonal demands and manufacturing the right supply of the latest clothing at the right time. Tech-savvy designer brands like Tukatech, Optitex, Accumark, and Reach are other leading players who have leveraged computer technology and robotics to achieve their production goals quickly and with higher efficacy. One such benchmark created by Tukatech for apparel manufacturing is described below.

Styku is a venture of Tukatech [11] that is an apparel-based software used for developing digital pattern making and converting the garment patterns into 3D applications like rendering a dress on the dummy to check the proper fit of the garment. It has become a revolution in the world of fashion, as it makes easier to understand the "fit" of the garment. In the right panel of Fig. 3.10, it is clearly depicted that data are combined to make accurate 3D renderings and animations, and also can be used to generate "heat maps" that show where a garment fits tightly (red on a heat map) and where it fits loosely (green) (Figure 3.10).

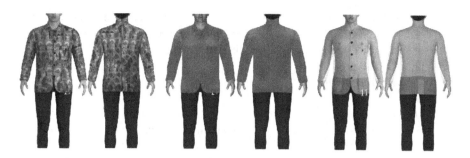

FIGURE 3.10: Example of a virtual fit [13]. a) Complete view; b) visualization of tightness of fit; c) transparent view.

3.8.2 MARKET OUTREACH PERSPECTIVES

- *Learning Market Demand*

  Now the world is at the cusp of the next industrial revolution, yet recent history is filled with tales of business companies as well as the industries that have committed errors in accurately determining total market demand. Understanding the law of supply and demand is crucial for retail stores, designers, and garment manufacturers to determine what products the customers will be happy to buy and at what price. Exploring the relationship between quantity and the minimum selling price of a product leverages price elasticity, and thus offers better decision-making capabilities.

  Adoption of analytics and computing tools by small and medium-sized businesses and big industries has soared in recent years for the construction of scalable, efficient, enterprise-grade design models. The application of digital software, robotics, 3D printing, wearable technology, big data, virtual reality, and AI has made it easy for fashion designers to implement the higher form of intelligence algorithms, contributing to a rise in the economy and standard of living. These algorithms identify and then recommend samples of apparel that could be used as an inspiration to design a new garment. Furthermore, some specific algorithms are also available to capture distinct attributes in order to augment the original style with more interesting and appealing patterns.

  Stitch fix is one such brand that uses its proprietary AI to analyze, identify, and generate unique fashion designs that are in high demand by the consumers of fashion and yet are missing from the market.

- *Building Sociocultural Context*

  Creating a sociocultural framework for garment construction accelerates the design process. There have been many huge leaps, fashion designers are often surprised to learn, that technology has made to ease their work. One such application of digital technology is in learning customer touch points through social media platforms like *Instagram* and *Pinterest*, which also forecasts the future of fully independent AI fashion designers at work. In order to assess and improve software-made proposals into fashion apparel of interest to shoppers, social media platforms are a boon to determine a brand's USP product.

- *Virtual Merchandising*
  Amazon
  One of the latest innovations in AI-powered visual style assistant is Amazon's Echo Look, which shows a deep understanding of fashion styles. Nevertheless, StyleSnap is another benchmark introduced by this online shopping giant that comes in-built into Amazon's mobile application and helps customers quickly search for clothes to purchase [1]. This facility allows users to upload an image of their interest, and the software works to match the look in the image and showcases similar related results available on the site.

- **Cloud Avatars for Smart Clothing**
  Smart clothing can make a significant difference in ensuring a healthy lifestyle. Wearing smart clothing suitable for various routine activities in suitable environments is often considered part of a healthy lifestyle. With research on the cloud-based approach to assist you in selecting the best outfit for a particular situation, visual avatars [3,4] are now available in the digital market who can dress you well and influence your personality.

Keep your comfort level high and adjust as the weather dictates. Immersive AI technology offers you a talking assistant for wardrobe selection based on your body type, target activity, environment, and the best-fit fabrics. The idea is of an app that allows you to watch yourself in the mirror, talking to a virtual style assistant that keeps on suggesting attire for some activity! It has become possible.

The recent research proposed by academic domain experts is a fast clothing simulation system that consists of four pillars: human body type, activity, environmental factors such as temperature and humidity, and the fabric of the apparel. The user just enters as much personalized information before selecting the garment, and the app will show several suggestions for the best attire to wear [6] (Figure 3.11).

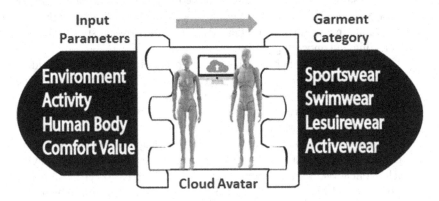

FIGURE 3.11: Cloud avatars for selection of garment categories.

3.9 FUTURE OF FASHION DESIGN

Clothing is the physical form which is considered as the second skin of human body and determines the utility of fashion design based on "consumer-oriented" strategies. The emerging demands of social attributes for a fashion trend and the lack of resources to meet consumer needs have created a need for designers to ponder upon the future of fashion design. In order to meet the global market demands, designers need to think deeply about the diversity background of people reflected in the essence of their clothing styles, and this needs to be blended with the concept of multi-angle, multi-paradigm, and multifunctional design. This section of the chapter reveals the evolving strategies for futuristic fashion design that utilizes innovative science and technology advancements to accommodate changing social trends in the modern age.

3.9.1 FASHION THINKING + DESIGN THINKING = CROSSOVER DESIGN

The origin of "crossover design" is implicit in the name itself and depicts the paradigm of cross-border cooperation. The meaning of crossover design is the collaboration of people from different demographic locations with the instincts and knowhow of their respective cross-border cultural heritages. This has become a kind of new design method and strategy across the fashion world. The formulation of this strategy is aimed at offering an open platform to designers where they can express their freedom of imagination by expanding the dimensions of fashion design and broaden their horizons by grasping the new standards of design process for product development that surpass stereotyped methodologies. Fashion Designers can consider this design strategy as an opportunity to interlace the coexistence of global styles and to implement trans-domain design methodologies that meet the need of the hour.

3.9.2 CONVERTING "IMPOSSIBLE" TO "POSSIBLE": SMART CLOTHING

The fast-paced advancements in science and technology development have reshaped the form of clothing design inspired from the fashion arts. With the support of high-tech tools, modern clothing has been transformed to achieve consumer satisfaction and has expanded the space for development of smart clothing in fashion. With this transformation, the clothing industry is able to realize the future sci-fi clothing which often exists in peoples' minds and in imaginative science fiction. The orientation of smart clothing drives traditional styling toward intelligent garment design. The social benefit of this strategy is that business players in the field of modernized clothing are receiving more requirements from the buyers' market. Fashion consumers are nowadays more fascinated with smart clothing that meets human functions in dealing with complex survival environments.

3.9.2.1 The Development of Innovative Fabrics

Gone are the days of fabric only being used for clothing purposes; now, new innovations in fiber properties and abilities to be constructed pave the way forward for the fashion industry, allowing it to be a dynamic one that is forever changing. In order to develop innovative fabrics, a broad space of collaboration should be provided between designers along with other disciplines like human body engineering, medicine, chemical technology, nanotechnology, biotechnology, optics, etc. in order to develop high-performance clothing like fireproof fabrics, stab-resistant and bulletproof fabrics, and antibacterial fabrics for medical applications.

3.9.2.2 The Combination of Electronic Products

Fashion is a sort of "encounter with time". The future designers belong to those who use the technologies of their time. But both technology and fashion embody the most fragile and ephemeral aspects of our culture, insofar as that what is cutting-edge today will be old tomorrow. Fashion designers have known for a long time that they are working with a fleeting material that will never be timeless. With the continuous innovation in technology, new territory has opened up for fashion design called smart fabrics, which respond not only to the wearer but also to actions. These are the fabrics with built-in high-tech electronic products such as microsensors and other new technologies that can provide added value to the wearer. The design of intelligent clothing has the ability to do many things that traditional fabrics cannot, including communicate, transform, conduct energy, and even grow. With the advent of continuous improvement in the function and design, these fabrics have been utilized in several advanced areas like the military, medical, industry, automotive, entertainment, and public sector domains.

3.9.3 NEW DESIGN PATTERNS: VIRTUAL INTERACTIVE DESIGN

The evolution of virtual interactive design in fashion technology has made fashion designers more creative, which reflects in works of art of high production and academic value. Digital and network technology makes designers more comfortable to present their designs immediately and satisfies modern people's psychological desire for convenience through virtual reality. With the help of CAD or CAM, designers can make out complicated patterns in a couple of minutes and receive approvals online that in turn save a lot of manpower and material resources. The concept of virtual interactive design for fashion clothing can be interpreted using a two-faceted view:

3.9.3.1 Augmented Clothing Design

One of the most exciting developments in immersive design technology is augmented clothing design (a.k.a. virtual clothing design) in which fabric can be simulated swiftly into several styles. This revolutionary technology in the design process has increased accuracy and production, shortened lead times, and expanded the capabilities of designers to outreach diverse consumers. Virtual design technology enables lively experience by its users to "see" the effects, "smell" the smells, "touch" the textures, and "hear" the sounds, while "absorbing" the impact of the software on their minds and lifestyle. This technological design has emerged as a brand-new clothing concept which renders a new way and process of fashion design. The prolific advantage of virtual clothing design is that it makes fashion designers comfortable in visualizing more variations in real time, no matter how complex or experimental they may be at bottom; that's the beauty of tapping "Ctrl+Z" and "Save As" on their machines.

Today, a lot of research institutions are succeeding in the virtual clothing design domain, like MIRA Lab at the University of Geneva, Switzerland, for creating its interactive environment of fashion design and simulation, and Toyobo of Kyoto, Japan, for its research in developing Dressing Sim digital fashion software. To name just two industry players: triMirror created a benchmark with the world's first real-time, one-of-a-kind virtual fitting solution, which not only enables consumers to experience using their actual designs and fabric characteristics as they try on real clothes, but also assists the designers to shape real-life clothes on their accurate virtual models in motion through visualization on web, desktop, and mobile platforms; and CLO Virtual Fashion LLC is actively producing a 3D technology environment in which consumers can select suitable clothing for themselves through the network as long as they continue uploading their body data.

3.9.3.2 Ultra-Dimensional Visual Clothing Design

Ultra-dimensional visual clothing design is another interesting novel concept of modern fashion design which exemplifies a hub of superficial apparel design networks. This design hub is used to dig for insights into people's psychology, sense of vision, and aesthetics by exploring pentagonal sectors of line, plane, solid, time, and space. The analysis of this network enriches designers' ability to visualize multidimensional design thinking. Designing apparel based on color, picture shape, and composition is now an obsolete trend replaced by humanizing design. The urge of humanizing design technology is the need to understand behavioral science, environmental science, and technology in transforming the costume design from two-dimensional objects to three-dimensional. The design of network clothing not only gives importance to the dressing effect of clothing and imagination, but also focuses on creation of visually comprehensive circular system of human, clothing, and environment. In brief, this concept interlaces spatial morphology with other psychological factors in clothing design.

3.9.4 Converging Fashion Design and Sustainability

The concept of Convergence between Fashion Design and Sustainability is to stylize a product organically with a hot new pursuit of fashion. The concern for sustainability is also relevant for the fashion industry, as it faces several challenges in terms of the large quantities of water and chemicals used in the different phases of the production of clothes and their effect on the environment. These devastating and abiding trends put human survival at stake against various global environmental disruptions, such as global warming, energy depletion, and water, air, and land pollution. In the pursuit of economic benefits, designers and technologists should pay more attention to the quality of life and realize the importance of ecological environmental health.

Possible solutions to severe problems related to the environment are as follows: First, to make a system of transparency that offers consumers confidence in the certification that accredits certain brands as ecologically sustainable; and Second, to implement a policy of "work on demand" that can drastically reduce the implicit environmental impact.

3.9.4.1   Materials for Green Fashion

Green fashion is an emerging and very exciting phase of sustainable movement for the designing of ecofriendly materials which also do not harm human skin. With a growing awareness among consumers, large numbers of clothing companies are transforming their business models to include sustainability procedures in order to improve their supply chain, reduce adverse environmental impact, and improve social conditions at the workplace. Despite the fact that we have not eliminated the use of harmful materials like polyester and nylon in the fashion industry, we can gradually drive a shift to the use of alternative materials that reduce post-consumer waste and the content of landfills, ultimately curbing pollution. In fact, the concept of green fashion has marked its presence in the process of production to achieve a "no-harm" objective.

Designing sustainable clothing requires congenial raw materials or green materials like organic cotton, bamboo fiber, natural silk, soybean fiber, milk fiber, corn fiber, and model and green regenerated fiber for synthesizing ecofriendly fabrics. In this regard, the re-engineering of textiles and synthetic materials has also proved a boon in responding to the needs of customers and in protecting the environment. Simultaneously, these new materials also compensate for the shortcomings of traditional green fibers by making the clothing more comfortable, breathable, and functional.

3.9.4.2   Revival of Traditional Crafts with Modern Technology

There is no question that the top priority is getting what is needed and where it is needed as fast as possible. This can be achieved by either relying on existing traditional craft workshops, or setting up a modern manufacturing space equipped with the latest technology for the production of weaving, knotting, embroidering, wrinkling, topstitching, pulling the net, beading, tie-dying, batik, bonding, stitching, and hollowing out. An exciting way for different industries is to use basic manufacturing methods for creating a wide array of engineering and durable components. The combination of traditional crafts and modern technology not only improves the efficiency of clothing production but also raises the economic and environmental value of research and development.

Leveraging modern technology [7] in the industry not only creates job opportunities for craftsmen and artisans but also has substantially contributed to the revival of sluggish traditional craft forms. The combination of traditional crafts and modern technology can be used to generate a new artistic effect, which is greatly reflected in the emergence of green design and the reuse of fabric. The modern seamless splicing, digital printing, and laser cutting technology can make up for some limitations in the traditional craft. Traditional handicrafts and different methods for secondary design of fabric and building new fashion spaces can be used to get richer fabric effects and make the clothing more fashionable. Most important is that fabric reconstruction is more in line with the requirements of environmental protection clothing.

3.9.4.3   Recreation of Green Textile

There has been a push to make water-soluble plastic clothing among British designers and scientists. The intent behind this innovative scheme is to educate people about the environment through clothing design patterns. Recently, a new avatar of green textile was presented by two researchers: Helen Dorian (London College of Fashion), and Tony Ryan (the University of Sheffield). The invention is the result of their search for a method to reduce discarded clothes that are sent to landfills, causing pollution and wastage.

Different kinds of plastic clothes made of biodegradable polyvinyl alcohol that can dissolve in water are in vogue. Reportedly, global researchers are engaged in studying the design approach for "contact reaction clothing". The study aims to propose a solution that makes use of washing and neutralizing the surface of the garment to make it pollution free. The team proposes that the large surface area of the clothes can otherwise be used to purify the air.

SUMMARY

The purpose of fashion design is not only to pursue fashion and aesthetics, but also to facilitate the understanding of design art that imbibes the essence of multicultural, intelligent digitization for the benefit of mankind. Metamorphosizing from a product-oriented to a people-oriented paradigm, the field of fashion design is leveraging the best of science and technology to meet the social-cultural, psychological, and physiological needs of the consumer world.

With better designing and customized solutions, there are multiple ways in which science and technology are impacting the global fashion industry. AI and machine learning tools are also helping, leading fashion brands to increase return on investments and to scale up their existing operations.

This chapter presents the ideal perspective of fashion design with new methods and benchmarks for the designers to harmonize the state of the art with a scientific and technological foundation for the benefit of environmental, industrial, and ideological shifts. Nevertheless, technology in the field of fashion is moving out from the palm of our hand and onto the sleeve of our shirt.

The demand of people for comfort and aesthetics has gradually achieved integrated production-green-health with the help of science and technology. Researchers and innovators in the fashion industry are helping out the designers and garment technologists not just in building up garments to be aesthetically strong but also in making it possible to encapsulate more functionalities like color change technology, phase change materials, and protective and smart wearable clothing. While these possibilities continue to evolve in various dimensions, a continuous stream of the untapped opportunities is pushing the fashion industry toward the new future.

REFERENCES

1. https://www.theverge.com/2019/6/5/18653967/amazon-fashion-ai-stylesnap-mobile-app-clothes-search
2. https://medium.com/@techpacker/13-must-have-apps-for-fashion-designers-85b1ab3e85fa
3. https://www.fastcompany.com/1750193/the-avatars-new-clothes
4. https://www.scmp.com/lifestyle/fashion-beauty/article/2159582/fashions-first-avatar-supermodel-could-mean-most-beautiful
5. https://www.computer.org/publications/tech-news/research/talking-fashion-avatar-mirror
6. https://technicaltextile.net/articles/spray-clothes-on-yourself-seamless-garments-5177
7. http://www.tpof-thepsychologyoffashion.com/industry-articles/fashion-art-or-science
8. https://vocal.media/01/7-modern-technologies-used-in-fashion-design
9. De Raeve, Alexandra, Cools, Joris, and Vasile, Simona, "3D body scanning as a valuable tool in a mass customization business model for the clothing industry", *Journal of Fashion Technology & Textile Engineering, Scitechnosol* 4 (2018), 009. doi:10.4172/2329-9568.S4-009
10. https://all3dp.com/2/3d-printed-fashion-the-state-of-the-art-in-2019/
11. https://techpacker.com/
12. Hein, A.M. *Daanen and Faculty of Behavioural and Movement Sciences*, Vrije Universiteit Amsterdam. www.tukatech.com
13. Seamless Knitwear: New technology ensures one-piece construction with minimal wastage, Apparel Resources. 2008. https://apparelresources.com/technology-news/manufacturing-tech/seamless-knitwear

4 Science in Textile Design

4.1 EVOLUTION OF TEXTILES

As a uniquely abstract art form, the textile segment provides a detailed illustration of how with the era of human civilization, humans absorbed knowledge from the environment and their experiences and became able to transfer this wisdom from one generation to the next. The timeline of textile development dates back to the origins of civilization and shows many landmarks in the growth of man's desire for artistic expression. The use of textiles for personal and ceremonial purposes has evolved through the centuries in various ways. The need for material has always been the mother of remarkable inventions in textile science and apparel design. Sourcing the fiber from flora and fauna, converting it to fabric, and producing material for clothing, ceremonial, and decorative purposes all reflect the journey of textile evolution for sustainable living. Nowadays, the textile industry is leveraging material science practices, and advancements in technology have enabled manufacturers to produce smart wearables with bio-friendly properties.

Since the commencement of the industrial era three decades ago, many revolutionary events have taken place within the textile arts, and textiles as a manual art have become a highly developed form of visual artwork. Historically, the invention of mechanics, powered tools, and information technology have transformed this far-reaching industry into a highly developed one, even more so in the past few years. The modern textile industry is well-known for its versatile business framework and its contributions to the global economy. In contrast, there have been many areas of textile manufacturing untouched by the mechanical revolution; for instance, handmade textile development remarkably still continues at the same level as 300 years ago, even in countries were machines are replacing manpower in other economic zones. Driven by the indispensable knowledge and skills of artisans, chemical science professionals, technical apprentices, designers, and visual merchandisers, the industry has achieved a state of tremendous mass production without a significant change in the core operations performed for converting fibers into useful textile substrates. Nevertheless, as notable progress continues, the requirement of manpower cannot be obviated due to workers' skillful knowledge and long-held experience in assessing a material on a qualitative basis for making high-performance textiles.

4.2 SCIENCE IN TEXTILES

Contemplating the role of science in textile design has stimulated great interest in both designers and scientists to collaborate and study a wide choice of polymeric fibers, examine their properties, and explore scientific design approaches to create innovative products that can improve the safety and comfort of consumers. The involvement of science in textiles is contained in the physical properties and principles of material selection to manufacture yarns and fabrics. Textile science employs factors including the development of metal and alloys for machinery, new production mechanisms, and power generation, benefiting a large number of services offered to the industry.

Science in textiles is majestic and has been distinguished by the incorporation of improved polymeric fibers and analysis-driven methods of production. To meet the particular demands of the fabric in different aspects of quality and processing, science is always integral to generating different varieties of fibers that exhibit specialized chemical and physical properties. As a consequence, the improved materials find applications in areas including but not limited to e-textiles, smart apparel, packaging films, engineering goods, and medical kits. This chapter reveals the vast involvement of science by studying the polymeric materials themselves, their properties, and how they should be handled to make them suitable for use in the development of ideal products.

4.3   MATERIAL SCIENCE: FROM FIBERS TO FABRICS

Material science considers quantitative and qualitative research for generating improved textiles or fabrics. Science in textile-related studies is occupied with multiple instruments, internal bonding patterns of materials, chemical processes, and the blending of different fibers to make substrates.

The science of materials in textiles deals with the techniques of exploring and analyzing the generalized structure of a textile fiber. Accelerated by the discovery of chemical laws and by polymerization experiments, the development in fabric studies and techniques has outpaced the performance of textile science in many research areas of industrial importance. With inundated datasets and new benchmarks in textile structures for quality improvement, the majority of fiber manufacturers are reaching out to industry players to deliberate over fiber properties that will enable them to keep up with the ever-changing market demands.

4.4   ESSENTIAL PROPERTIES OF TEXTILES

Textile materials combine a number of distinctive properties to make textile structures, typified by woven or knitted fabrics, that are assemblies of filaments held together by purely geometrical constraints in such a way that the basic strength and flexibility of the filaments are preserved and utilized to the maximum advantage in the finished structure.

1. Comfort properties:
   The flexibility is a basic property of the structural element, that is, the fiber or filament; it increases as the filament diameter is decreased. For clothing purposes, comfort and warmth are the primary considerations; these are determined by the water-absorbing properties of the filament material, and by the transfer of heat (mainly by convection) through the fabric structure.

The absorption or release of water from the material plays an important role in the adjustment of the body temperature to variations in the temperature and humidity of the surrounding atmosphere. For undergarments, a structure of low density (that is, one that contains a large proportion of air space) is desirable. Traditional materials such as wool and cotton, which are made up of relatively short fibers, automatically produce an open type of yarn structure owing to the presence of loose fiber ends and the general irregularity (crimp) of the filaments. These materials are also capable of absorbing large amounts of water (35% by weight in the case of wool and 20% for cotton). In this respect both cotton and wool satisfy the basic requirements for comfort. Equally satisfactory is viscose rayon produced from wood pulp – one of the earliest and still one of the most extensively used of the man-made fibers – which is normally supplied in the form of "staple" (short-length) fibers prepared by chopping up the continuously extruded filaments.

2. Physical properties:

These are the properties useful for determining resistance to tearing and abrasion, desirable draping and handling properties, resistance to creasing, and stability of form (including retention of pleats or folds). These are complex properties, dependent on the type of fabric structure as well as on the inherent properties of the filament material.

3. Chemical properties:

These are also involved in dyeing, laundering, and other processing treatments.

Reaching to the conclusive point, new developments in any sort of textile are dependent on the collaboration of individuals from the verticals of the science and design domains, e.g., physicists, biologists, chemists, and design engineers.

4.5   TEXTILE SCIENCE LABORATORY EXPERIMENTS

Textile science is essentially the "why" and know-how of the current state of any textile substrate. Textile science labs lay the foundation of analyzing the characteristic mechanical behavior of textile fibers, such as loading rate, creep effect, and stress conditions dependent on time and temperature as well as the relative humidity to which the fiber has been exposed. Textile laboratories also fulfill the requirement of examining the chemistry of the coloring and blending of fibers to enhance the appearance of a given textile material. Textile science labs aid textile chemistry scientists in developing basic dyeing blueprints that tend to raise the level of understanding and enable textile designers to create interesting and artistic patterns, despite lacking scientific knowledge.

Color is another very important ornamentation and also the most-sought scientific textile application. Deploying a color grade to textile surfaces requires the human ability to recognize color differences. Also, designers with a great understanding of aesthetics give great attention to the variety of colors and their specifications while choosing the physiological aspect of textile development. Making use of textile laboratories, a textile colorist succeeds in making a vast array of colored textile samples with precision, and in avoiding color mismatches for same selection by observation under different sources of illumination. Leveraging science in textile coloring demonstrates progressive results achieved through knowledge of the physics of color, by color specification, and by using equipment like spectrophotometers and colorimeters for color analysis and control.

A textile science laboratory offers various advantages, including the following:

1. Offers understanding of the complex realm of the stress-strain behavior of fiber, yarn, and fabric in tension, torsion, compression, impact, shear, and complex combinations of these effects for various lengths of time and under varying atmospheric and geometrical structural conditions.
2. Provides new horizons by the use of electronics in physical measurements like stress transducers combined with high-speed recorders to make rapid and precise measurements of these physical quantities.
3. Aids in separation of elasticity components under stress/strain, with acoustical pulse techniques for the measurement of sonic modulus.
4. Helps in development of the science of premixed dyestuffs on a large scale, beyond a small selection of basic colors, to reproduce anything from an old "dark" look to a new shade of "surprising bright".
5. Allows designing infrared detection instruments that extend the principle of camouflage outside the visible color spectrum.

Experimental activities in laboratories add to the knowledge about new scientific techniques for designing new fibers with predictable properties, understanding traditional and modern textiles, and solving problems quantitatively.

4.6 MANUFACTURING PROCESS AND INTERNAL GEOMETRY OF TEXTILES

Textile design is the process of planning and producing a fabric's appearance and structure; in this process, fibers are only the first element of structure to consider as a textile's raw material. Afterward, they are twisted into continuous yarns. The yarns are formed into fabrics through different modes: woven, knitted, and nonwoven, as depicted in Figure 4.1. Science plays a major role in design development of textiles either by selection of particular fibers or by mixing two different fibers to set the material for yarn as per the design requirement.

Textile is not just a jumbled business of cloth and colors loosely thrown together but a well-organized industry driven by intensive market research and the intervention of science at different stages of textile production, including spinning, weaving, printing, dyeing, finishing, pattern-making, branding, and labeling, to enhance the aesthetic and value addition of the fabrics. A fiber is a small, short piece of hair of substantial length and diameter. A filament yarn is a long strand of a single substance. In textile yarn, individual fibers or filaments are tied together to make threads. Textile yarn can be made with natural fibers from substances such as wool from sheep, silk from silkworms, or cotton and linen from plants. It can also be made

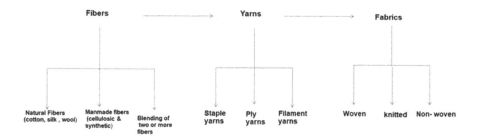

FIGURE 4.1: Geometry of textiles.

with synthetic or man-made fibers created from a variety of substances like nylon, acrylic, and polyester.

The process of making yarn is called *spinning.* Yarn can be spun by machine or by hand. Yarn used for weaving tends to have a tight twist, smooth surface, and lots of lengthwise strength. Yarn for knitting has a looser twist. Many specific production methods result in an endless variety of yarn. Textile yarn is made in a global industry that involves many specialized technical terms. *Weaving* is a process used to create fabric by interlacing the two sets of threads, namely, warp and weft. The process begins with the warp threads, which are stretched tight on a frame and run along the fabric's intended length. Weft threads are laced over and under and run perpendicularly to the warp threads.

4.7 SCIENTIFIC THEORY OF TEXTILE DESIGN

Textile performance can be enhanced by understanding the scientific background of raw materials pertaining to their physical and chemical properties. This also means that a precise comprehension of relationships between *fibers in yarns* and *yarns in fabrics* is the only tool for state-of-the-art textile performance analysis. In terms of the role of science in textile studies along with modern scientific advancements in the realms in electronics and nuclear physics, there still a lack of challenging research opportunities as compared to the fast expansion of scientific approaches in use by the textile design industry.

4.8 PHYSIOLOGICAL THEORY OF TEXTILE SCIENCE

The driving principle of textile design appears based more on the "stylish" say of designers than on "both stylish and comfortable". In the age of fashion growing roots in almost every region, aesthetics has attracted more designers to create masterpieces and consumers to buy eye-catching products. Researchers have elevated interest in exploring the relationship between textile materials, their properties, and their use by a human as an outfit. Thus, the roles of physicists, physiologists, and design engineers are combined to create a viable product. Whereas the physicist works on tracing the relationship between atmospheric stresses and human comfort, the physiologist and the engineer further tap the underlying principles governing the use of textiles as a protective shield against danger and harsh climates.

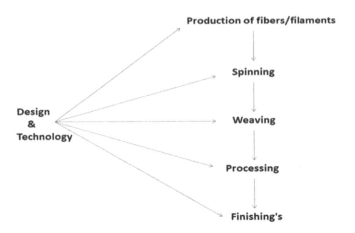

FIGURE 4.2:   Role of design and technology in textile manufacturing.

4.9   SCIENCE AND TECHNOLOGY IN TEXTILE DESIGN

Innovation-driven design methodology has been increasingly adopted in benchmarking product portfolios in the industry and also in monetizing the untapped potential of science and technology in order to add to product efficiency. With the emerging novel digitization techniques, the textile industry is shifting to the paradigm of automated technology, which results in quick development and delivery of finished products with less time to market along with better outreach.

Textile Science Research integrates development, collection, selection, and utilization of textile design methods in order to provide a rich experience of enhanced comfort and safety for textile consumers. Behind every innovation is the body of knowledge, design science research, and its methodologies, improving the performance of the existing technology to achieve automation.

Design research pays off a great deal in driving innovation in product design through understanding historical milestones and trending breakthrough tools utilized for creating a benchmark solution. The following section sheds light on how science in combination with design and technology is leaving footprints by increasing production as well as design variants matching the market needs for fibers, yarns, and fabrics (Figure 4.2).

4.10   DESIGN AND TECHNOLOGY IN TEXTILE MANUFACTURING

4.10.1   PRODUCTION OF FIBERS AND FILAMENTS

A fiber is generally hundreds of times longer than it is wide, giving it a hair-like appearance. Fibers can define almost every aspect of a finished textile. Fibers are incredibly important to textile production, but not all fibers are suitable for textiles. Textile fibers are those which have properties that allow them to be spun into yarn or directly made into fabric. This means they need to be strong enough to hold their shape, flexible enough to be shaped into a fabric or yarn, elastic enough to stretch, and durable enough to last. Textile fibers also have to be a minimum of 5 millimeters in length. Shorter fibers cannot be spun together. Synthetic fibers are made of polymers. The word "polymer" refers to a chemical substance composed of molecules that form long repeating chains, a characteristic that is useful in synthetic fibers/filaments. Synthetic fibers begin as chemicals, often derived from products like coal and petroleum. Depending on the type of fabric, these chemicals are combined with acids and alcohol, sometimes heated, and then extruded. Extrusion is a manufacturing process where a chemical substance is pushed through a die or nozzle to form long threads. So, all textiles are made of fibers, but not all fibers can be used to make textiles. For example, cotton plants contain fibers that are strong and pliant enough to be spun into yarn.

1. Designing fibers/filaments:
   Fibers are generally designed based on their properties; fibers designed in the laboratory can behave similarly to any number of fibers which originate from nature. Therefore, it's important to consider that not all textile fibers are created equal. Each fiber contains different qualities and will result in a different textile. They are differentiated by their properties: some retain heat better than others; some hold dye very well; some are more durable; and some are more comfortable. An important section of the textile industry is concerned with industrial yarns and fabrics, for example, tire cords, conveyor belts, parachute materials, and so on. For these, high impact strength and abrasion resistance are primary considerations. In such cases, the most effective results are obtainable by blending two or more different components. Thus polyester fibers, which possess outstanding tensile strength but do not absorb water to a significant extent, are frequently blended with wool or cotton to produce a clothing material combining high strength and resistance to abrasion with acceptable hygroscopic properties.

2. Technology in production of fibers/filaments:
   In most sectors of textile manufacturing, technology is one of the major keys to quality improvement and cost competitiveness; a good example can be seen in the collection of cotton, ginning, and its various stages of processing. In addition to its use in the production of natural fibers, technology is widely used in the manufacturing of fiber/filaments artificially, with much attention toward product uniformity and adherence to meet quality standards for fiber diameter monitoring, temperature and tension control, and monitoring of the solution properties of the polymer. With the advent of the latest technology, SEM images are used to learn about the complete DNA of fiber/filament [1].

   Nowadays, there is a race to make finer filaments, which bring revolutionary changes in producing vast varieties of fiber/filaments from cotton to man-made fiber and from microfiber to nano-fiber, which itself necessitates changes from traditional machinery to the latest machinery for spinning the fiber/filaments into yarn. This filament is a polymer, which is made up of monomers that are proteins (which in turn are polymers of amino acids). Bundles of such filaments comprise a cytoskeleton which fills several important roles:

- It supports the cell and maintains its shape.
- It holds cell organelles and other particles in position within the cell.
- It moves organelles and other particles around within the cell.
- It is involved with movements of the cytoplasm called cytoplasmic streaming.
- It interacts with extracellular structures, helping anchor the cell in place.

There are three components of the eukaryotic cytoskeleton: microfilaments (smallest diameter), intermediate filaments, and microtubules (largest diameter). Three highly visible and important structural components of the cytoskeleton are detailed in Figure 4.3. Specific stains were used to visualize them in a single cell. These structures maintain and reinforce cell shape and contribute to cell movement.

1. Microfilaments

Microfilaments (Figure 4.3) are usually made of actin and appear in bundles. Each filament is about 7 nm in diameter and up to several micrometers long. Microfilaments have two major roles:

- They help the entire cell or parts of the cell to move.
- They determine and stabilize cell shape.

Microfilaments are assembled from actin monomers that attach to the filament at one end (the "plus end") and detach at the other (the "minus end"). In an intact filament, assembly and detachment are in equilibrium. But sometimes the filaments can shorten (more detachment) or lengthen (more assembly):

$$\text{Actin polymer (filament)} \rightleftharpoons \text{Actin monomers}$$

This property of dynamic instability is a hallmark of the cytoskeleton. Portions of it can be made and broken down rather quickly, depending on

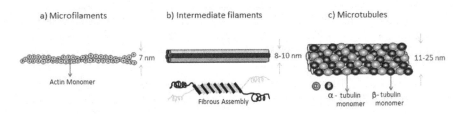

a) Microfilaments   b) Intermediate filaments   c) Microtubules

7 nm   8-10 nm   11-25 nm

Actin Monomer

Fibrous Assembly   α - tubulin   β - tubulin
                      monomer   monomer

FIGURE 4.3: The cytoskeleton (courtesy of Vic Small, Austrian Academy of Sciences, Salzburg, Austria [2]).

cell function. Actin-associated proteins work at both ends of the filament to catalyze assembly and disassembly.

2. Intermediate filaments

There are at least 50 different kinds of intermediate filaments that are diverse and stable in nature (Figure 4.3b), many of them specific to just a few cell types. They generally fall into six molecular classes based on amino acid sequence. One of these classes consists of the fibrous keratin proteins, which are also found in hair and fingernails. The intermediate filaments are tough, ropelike protein assemblages of 8–12 nm in diameter. Intermediate filaments are more permanent than the other two types of filaments and do not show dynamic instability. Intermediate filaments have two major structural functions:

- They anchor cell structures in place. In some cells, intermediate filaments radiate from the nuclear envelope and help maintain the positions of the nucleus and other organelles in the cell.
- They resist tension. For example, they maintain rigidity in body surface tissues by extending through the cytoplasm and connecting specialized membrane structures called desmosomes.

3. Microtubules

Regarded as the thickest elements of the cytoskeleton, microtubules (Figure 4.3c) are long, hollow, nonbranched cylinders about 25 nm in diameter and up to several micrometers long. Microtubules have two roles:

- They form a rigid internal skeleton for some cells or cell regions.
- They act as a framework along which motor proteins can move structures within the cell.

Microtubules are assembled from dimers of the protein tubulin. The dimers consist of one molecule each of α-tubulin and β-tubulin. Thirteen chains of tubulin dimers surround the hollow microtubule. Like microfilaments, microtubules show dynamic instability, with plus and minus ends and associated proteins.

$$\text{Microtubule} \rightleftharpoons \text{tubulin dimers}$$

Tubulin polymerization results in a rigid structure, and tubulin de-polymerization leads to its collapse. Microtubules often form an interior skeleton for projections that come out of the cell membrane, such as cilia and flagella [2].

Today, it is possible to find commercial examples for spin-draw-wind, spin-draw-warp, and spin-draw textile processes. These technologies place a different emphasis on material handling requirements; robotic technologies for package doffing and transport are increasingly available and yet because of the linking of processes, need be placed only at critical points in the overall process. The emphasis on flexible manufacturing, even in the fiber industry, has led to the development by some fiber producers of robotic techniques for the rapid change and replacement of spin packs and spin-nerets. In these examples, robots are called upon to do what humans cannot do – change hot parts before they have cooled.

4.10.2 DESIGN AND TECHNOLOGY IN SPINNING

Spinning is the process of converting the textile fibers/filaments and making them into yarn by twisting or other means of binding together the fibers and/or filaments. The term "spinning" can be associated with "fiber/filament" or "yarn" spinning. In conventional methods of spinning, natural fibers are into yarn by hand; nonconventional methods of spinning derive from more recent technologies and can enhance the appearance and quality parameters of yarn. Short, natural hairs that come from plants like cotton and animals like sheep are called staple fibers, and fibers that are long continuous single strands of more than a km in length are called filaments. Most filaments are synthetic or man-made materials, like polyester and nylon; however, silk is considered a natural filament. Adopting different methods of spinning made it possible to create new types of yarn from different fibers and synthetic filaments. In fiber/filament spinning, continuous filaments are extruded through a spinneret from fiber-forming polymers; these filaments can later be converted into staple fibers for staple yarns or can be the first step in producing filament yarns.

4.10.2.1 Designing in Yarns

Designers' foresight, imagination, innate sense of style, and desire to create innovative yarns has catalyzed various methods to produce state-of-the-art aesthetics and performance characteristics that may influence the look, feel, texture, and performance of the fabric:

1. *Fiber Blending*: The design of any fabric can be enhanced using a blend of fibers. There are various ways to enhance the aesthetic appeal of fabric:
   a. *Using Multi-colored Fibers*: Earlier, single-colored fiber was used to create yarn, which provided a single-colored fabric. Nowadays, using the latest technologies, multicolored fibers are spun together to make yarn, which enhances the aesthetic appearance of fabric. To produce multicolored fibers, we can use a single fiber in different colors or different fibers in a single color. Different fibers in a single color make a shade difference in yarn formation, which will in turn produce a yarn strand with multiple shades.
   b. *Using Different Content of Fibers*: Earlier, only a single type of fiber was used to create the yarn. However, using new techniques, different types of fibers can be spun together to make a yarn strand which will in turn enhance the aesthetic appeal of the fabric. For example, we can use pastel shades of cellulosic fibers and dark shades of man-made fibers

to make a strand which will ultimately enhance the aesthetic as well as functional property of the yarn.

2. *Physical and Chemical Modification*: There are a number of ways which will create fiber with modified physical and chemical properties. Here are a few:

   a. *Physical Properties*: Earlier, single-diameter fiber was used to create yarn. With advanced technologies, fibers of different diameters can be used to make a textured yarn. We can change fiber crimp and length as well to create innovative yarns which will in turn produce highly aesthetically improved fabric.

   b. *Chemical Properties*: By changing the softness of fibers, we can make yarn strands of differing softness, which provides smooth and rough textured yarns. Various modifications can be used to create these innovative yarn strands such as shrink resistance and tactile properties.

3. Adoption of Nonconventional Methods: Nowadays, conventional methods of producing a yarn have been replaced by nonconventional methods in order to increase the production and quality of the fabric. Various nonconventional methods are used, such as vortex spinning, microfibers, and splittable fibers.

4. Classification of Yarns: Yarn can be designed or produced by changing the assembly of substantial length and relatively small cross-section of fibers and/or filaments with or without twist, which can occur in the following forms:

   a. *Staple-Spun Yarns*: A staple-spun yarn is a linear assembly of many fibers in the cross-section and along the length, held together usually by the insertion of twists to form a continuous strand, small in diameter but of any specified length.

   b. *Filament Yarns*: Filament yarn is an assembly of continuous filaments, with the exception of monofilament yarn. The spinning of continuous man-made filaments follows the basic principle of pushing a viscous polymer dope through a number of fine holes (spinnerets) to produce

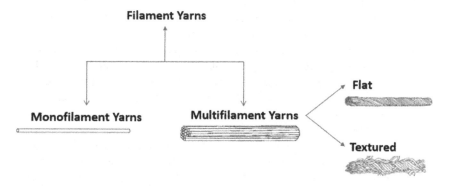

FIGURE 4.4:   Classification of filament yarns.

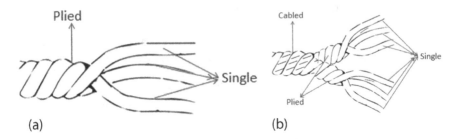

FIGURE 4.5:   a) Plied yarn; b) cabled yarn.

a slender stream of polymer, which is solidified by different means, attenuated, and finally collected on a suitable package. The two key types of filament yarns are *monofilament yarn* and *multifilament yarn* (Figure 4.4).

A single-hole spinneret produces a monofilament yarn, and a multi-hole spinneret produces a multifilament yarn. The multifilament yarns can be further categorized into flat and textured yarns.

c. *Composite Yarns*: Composite yarns are a combination of staple fibers and filament(s). The two main configurations include either a filament core covered by staple fibers (core spinning) or staple fibers wrapped by a filament (wrap spinning). In some instances, more than one filament can be present in the core, or the core can be dissolved in a postprocess to produce hollow yarns.

d. *Plied and Cabled Yarns*: A certain number of yarns are brought together with a slight twist (considered as plied yarns); in cabled yarns, a defined number of plied yarns are brought together with a slight twist. In contrast to single yarn, plied yarn is more stable, rounder, less hairy, more uniform, stronger, and more expensive. However, to meet product demands, it may be necessary to twist two or more single yarns to produce plied yarns (Figure 4.5a). The given twist is usually in the opposite direction to twist in the component yarn, to form a torque balanced structure. Cabled yarn is an advanced technology used to produce new high-performance types of yarns having new combinations of unique properties [3]. Cabled yarns (Figure 4.5b) can be produced by twisting two or more plied yarns together, usually for technical applications. For example, some sewing threads are cabled to attain maximum strength in conjunction with reduced irregularity, stretch, and liveliness. The plied and cabled yarns are also known as folded and corded yarns, respectively.

e. *Metallic Yarns*: A metallic yarn is a narrow strip of material, such as paper, plastic film, or metal foil, with or without twist, intended for use in a textile construction. These metal filaments had inherent shortcomings which restricted their use. They were expensive to produce; they tended to be inflexible and stiff; the ribbon-like cross-section provided cutting edges that made for a harsh, rough handle; they were troublesome to knit or weave, and they had only a limited resistance to abrasion. Despite these shortcomings, the metallic ribbon filament has remained in use for decorative purposes right up to the present day.

4.10.2.2 Technologies of Spinning for Textile Production

A range of spinning techniques was developed during the nineteenth and twentieth centuries, all with different characteristics in terms of the types of fiber that could be spun, the process economics, and the yarn's properties and applications. The main types of spinning systems are as follows:

1. Mechanical Methods of Spinning
    a. *Ring Spinning:* Ring spinning is the most common industrial method for short-staple fibers; in this method, twist is inserted in a yarn by using a revolving traveler. The majority of staple yarn manufactured worldwide is produced either by ring spinning (70%) or by rotor spinning (23%), with ring spinning dominating because of the superior quality (particularly of strength and evenness) in yarns produced using this method. Besides the good points of ring spinning, certain drawbacks limit the use of ring spinning due to the additional requirement of a roving passage and greater rotational speed of the package compared to the traveler. These make it an expensive method for yarn production since its inception in the conversion of staple fibers into staple yarns, as has been shown in Figure 4.6:

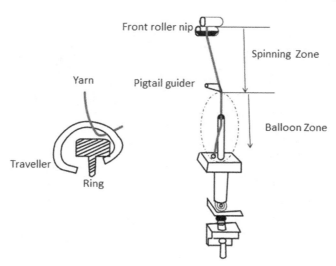

FIGURE 4.6:   Diagram of ring spinning [4].

b. *Rotor Spinning*: Rotor spinning is a technology of open-end spinning and the second most widely used for staple spinning after ring spinning. It uses a rotor (a high-speed centrifuge) to collect individual fibers for the production of yarn. The mechanism of this spinning is to open the individual fibers by opening the roller in their input sliver; then, air suction is used to collect and transport the opened fibers through the transportation tube. The narrowing of the tube helps in aligning the fibers, where centrifugal force is generated by the high-speed rotor and the fibers are collected in the rotor grove and form a ribbon around the periphery of the rotor.

By the action of the centrifugal force, seed yarn is introduced through the navel and tube and attaches to the fiber ribbon, where seed yarn rotation takes place, which leads to the application of one twist in the yarn by rotating each turn of the rotor. Rotor spinning offers high-production rates, but the limit is determined by the maximum rotor speed (around 175,000 rpm). The technology is suitable for short-staple spinning, as long-staple fibers would require low-production large rotors [5]. The applications of rotor-spun yarn in the clothing sector include denim fabrics, trousers, sportswear, shirts, blouses, and underwear (Figure 4.7).

c. *Friction Spinning*: Friction spinning is an open-end spinning in which the yarn formation takes place with the aid of frictional forces in the spinning zone. It uses the outer surface of two rotating rollers to

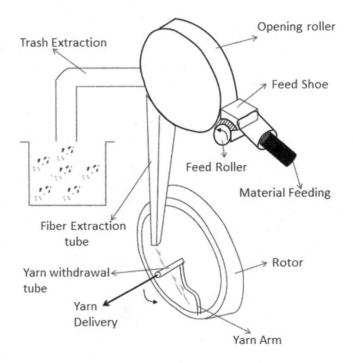

FIGURE 4.7: Diagram of rotor spinning [6].

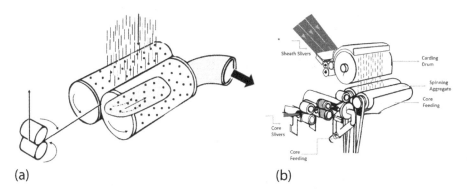

FIGURE 4.8:   a) Principle of open-end friction spinning; b) principle of false twist friction spinning [7].

collect and twist individual fibers into a yarn. In the open-end system (Figure 4.8), at least one of the rollers is perforated so that air can be drawn through its surface to facilitate fiber collection. The twisting occurs near the nip of the rollers and, because of the relatively large difference between the yam and roller diameters, high yarn rotational speeds are achieved by the friction between the roller surface and the yarns. The only commercially successful spinning systems are DREF 2000 (true twist in open-end form by mechanical means) and DREF 3000 (surface fiber wrapping around false twisted strands of fibers).

Friction spinning produces coarse yarns with low strength, owing to poor fiber configuration. The main area of application is sports and leisure clothing, protective textiles, outerwear, and technical textiles. It is also suitable for spinning recycled fibers [7].

2. Experimental Techniques in Fiber/Filament Production

The conversion of polymers into fibers/filaments through extrusion is called spinning, a technique used in the production of man-made fibers. There are several advancements for the production of filaments like melt, wet, dry, and gel spinning. The fiber-forming components, generally called polymers before being spun, must be converted into a fluid state and then forced through the spinneret, where the polymer cools to a rubbery state, and then a solidified state.

a.   *Melt Spinning*: This method is widely used for the production of polyester and nylon, where fiber-forming thermoplastic polymers are melted and forced through spinnerets. The melt solidifies immediately after issuing from the jets and so forms filaments, which pass through a cooling chamber in which a cold air current is swept across the filaments. The spinning speed is approximately 1200 meters per minute. Designing of filaments generally depends on altering the geometry of the spinneret holes from which the spinning dope is extruded (e.g., Coolmax from polyester). This method allows bi-component spinning by mixing two different components of polymers, which are extruded into single filaments in different arrangements, e.g., microfibers (Figure 4.9).

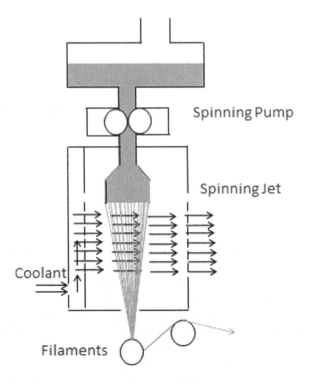

FIGURE 4.9: Schematic description of melt spinning [7].

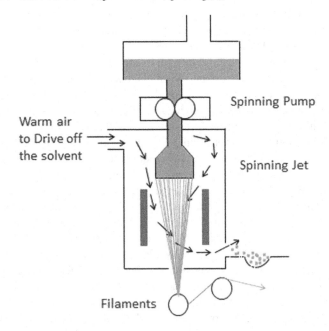

FIGURE 4.10: Schematic description of dry spinning [7].

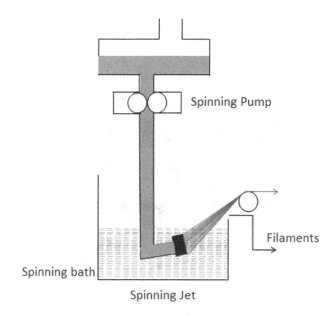

FIGURE 4.11: Schematic description of wet spinning [7].

b. *Dry Spinning*: This technique is used for the production of acetate rayon and acrylic fibers. In this method, fiber-forming materials are dissolved in a volatile solvent and then the solvent is evaporated with the help of hot air to solidify the filament (Figure 4.10).

c. *Wet Spinning*: This technique is used for the production of viscose, algi-nate acrylic, and para-aramid filaments/fibers like Kevlar, and Twaron, where the extruded filaments are coagulated in a nonsolvent liquid bath (Figure 4.11). If there is an air gap before filaments enter the coagulant bath, this is considered as a variant of wet spinning called dry-jet wet spinning.

d. *Gel Spinning*: In this technique, a highly concentrated spinning dope is extruded for the production of high-performance ultra-high-molecular-weight polyethylene that contributes to attaining the desired physical and mechanical properties of filaments. The polymer here is partially liquid or in a gel state during extrusion, resulting in strong inter-chain forces and hook-ups with a high degree of orientation relative to each other, which significantly increases the tensile strength of the fibers. This process can also be described as dry-wet spinning since the filaments first pass through air and then are cooled further in a liquid bath.

4.10.3   DESIGN AND TECHNOLOGY IN WEAVING

Weaving is the process of creating a textile by interlacing two types of threads called a warp, which runs vertically, and the weft, which runs horizontally. It's an old art form and one that only needs simple equipment. Most important among those is

Interlacement of warp & weft

Total - 16 Interlacement points

FIGURE 4.12: Interlacement of warp and weft.

a loom, basically a wooden frame that holds the warp threads in tension so weft threads can be worked over and under them in horizontal rows.

Weavers create cloth on a mechanical device called a loom. A loom is a machine that interleaves two sets of perpendicular threads. In one common format, one set of threads extends directly away from your body, and the others run left-to-right. Figure 4.12 shows a simplified loom. The red fibers that go out away from us are called the warp threads, and they're strung and held taut through mechanical devices called shafts. The blue horizontal fibers are called the weft threads, and they're strung and held taut through devices called treadles.

4.10.3.1 Designing in Weaving

Fabric design is a creative process with technical implications that require designers to combine a sense of visual appeal with a strong knowledge of material extraction and fabric production machinery. This section offers learning about methods and techniques of design in weaving that complements both the fundamental understanding of woven fabric design and structure and the objective evaluation of color in woven fabric arrangements (Figure 4.13).

1. Fundamental Weaving Techniques:
    Three fundamental types of basic weaves exist – the plain, satin, and twill – which give rise to endless variations and combinations. The most basic type of weave is a plain weave (Figure 4.14a), in which the weft passes over and under every warp thread. To make it, start on the left side, take a weft

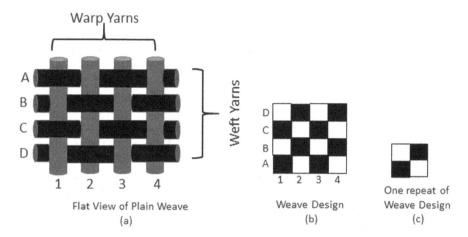

FIGURE 4.13:   Plain weave.

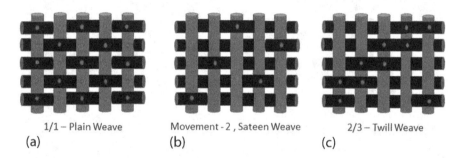

FIGURE 4.14:   a) Plain weave; b) satin weave; c) twill weave.

thread, weave it over the first warp thread and then go under the second. Repeat this process until you get to the last warp thread. Then begin another row with the weft thread, this time working from the right until you return to the left edge. Continue this process until the desired length of fabric is reached.

To create a **twill weave**, take the weft thread and pass it over two warp threads, then go under one warp thread, and then passes over two more warps Figure 4.14c. Again, start on the left, work to the right, and when you reach the right edge, begin the process over again, this time working right to left. Weaving in this manner creates a pattern of diagonal lines in the fabric and a fabric with two sides that look different.

Another common pattern is the **satin weave**. In this style, the weft threads pass over several warp threads before they pass under one (Figure 4.14b). Weavers stagger the interlacing so that no two overlaps are adjacent. Typically the warp threads are thin, and the weft threads are thicker than warp.

2. Color and Weave Relationship:

In order to create a striking pattern on the fabric, one must understand the relationship between colors and weave. The fundamental structure of weave along with the versatile combinations of colored warp and weft threads results in incredible designs in the fabrics. Figure 4.15a and Figure 4.15b are a few examples where numerous mixtures of colors to produce other colors can be obtained from a few colors of the warp and weft yarns through proper weave interlacing. It is clearly evident in Figure 4.15a, where color simulation is produced from pink warp yarns and blue weft yarns interlaced using a plain weave structure (see Figure 4.14a). The color simulation of Figure 4.15b is produced from pink warp yarns and blue weft yarns woven in sateen. This concept of a striking pattern can be created for jacquard design Figure 4.15c) by using the interlacement of a few colored warp and weft threads.

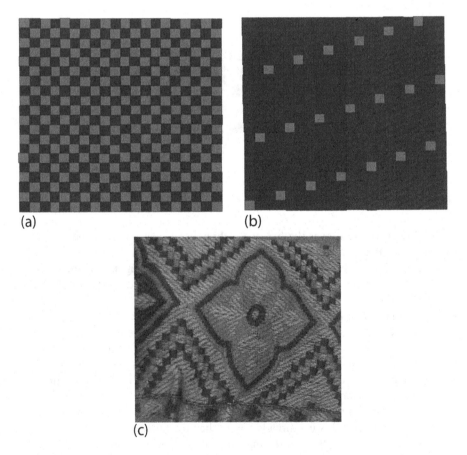

(a)   (b)

(c)

FIGURE 4.15:   a) Color simulation of the plain weave; b) CS of the sateen weave; c) CS of jacquard fabric.

4.10.3.2   Technology in Weaving

The textile industry is one of the largest in the world, employing millions of people globally. Textile materials are made from fibers, yarns, and fabrics. The earlier textile industry's focus was on the end product, but science and technology have brought a new dimension to the textile industry, including creating automatic machines for higher production, digital designing for weaving, and in-depth understanding of color science along with its visualization of woven fabrics and applications of CAD at several stages. The combination of science and technology provides manufacturers with a larger share of this market and greater profits.

1. **Weaving Using Digital Design**

   The utilization of computers at several stages of textile designing and weaving has transformed the entire thought process from the initial artwork to final production. CAD systems operate series of basic steps for woven design, which are as follows:

   - *Digitizing the Artwork*: The first step is digitizing the artwork, which allows the designer to see the artwork on a computer monitor by scanning the original piece or creating a design using the CAD system drawing tools directly.
   - *Fabric Designing by CAD*: The second step is fabric designing, in which the designer can concentrate on the finer aspects of their creations without any hindrance of limitations. The advent of CAD provides a comprehensive and efficient user interface that allows users to specify, construct, visualize, and execute designs in real time; in addition, automatically inserted weave structures are now convenient at specific areas of the fabric and can add desirable warp and weft structures to textile designs and integrate colors as per your loom, as shown in Figure 4.16:

     Figure 4.16 illustrates the design of fabric and fabric weave using Arahpaint and Arahweave jacquard software. Being digital, software such as Arahne provides a complete view of the technological implications of textile design and fabric weave imprints (Figure 4.17) without requiring any actual production. The software is also a promising tool for designers not only to create multiple weaves of jacquard but also to apply desired simulations (Figure 4.18) on the products to check the aesthetics.
   - *Weave Allocation*: The third step is weave allocation, in which information from the artwork image can be converted into a woven fabric. There is software that provides for creation of different design effects within the same draft order while changing the lifting plan and controls for creating complex weave structures by combining weaves, as shown in Figure 4.17.
   - *Simulation View*: This part of the program helps the designer to see a simulation of the final fabric on the display monitor [9]. By looking at

(a)

(b)

FIGURE 4.16:   Fabric designing by CAD software.

the preview in Arahne, the designer can easily modify the design, and can change the weaves to recolor the design as required (Figure 4.18).

By looking at the preview, the designer can easily modify the design and can change the weaves to recolor the design as required. All these developments have greatly increased the ease of woven fabric designing. It is now possible to perform the entire process on a personal computer, and then transfer the ready-to-weave file (electronic punch-card

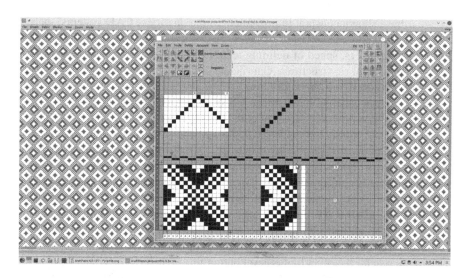

FIGURE 4.17: Weave allocation [8].

FIGURE 4.18: Simulation view [8].

file) via the Internet, direct to the dobby or jacquard controller at the loom, or to some interim storage area. There is also the possibility of seeing the resultant fabric on a computer monitor that gives a visualization of real fabric prior to weaving. Many CAD companies (UVOD, Fractal Graphics, Yxendis, ScotWeave, EAT, NedGraphics, Pointcarré, Mucad, InformaticalTextil, Booria CAD/CAM systems, Arahne Textile Designing Software [includes Weave, Paint, Jacquard, Drape, etc.])

have shown constant improvement in the quality of CAD systems, such as easy-to-use software modules, flexibility in changing constructional parameters, speed of defining technical data, and enhanced visualization of fabric structures (Gabrijelcic 2004, Seyam 2004).

2. Color Visualizing through CAD

Color curve generation and image processing provide opportunities for additional improvements in areas of collaborative color development, color marketing, and multi-step color prediction. At the same time, there are other aspects of imaging technology that have strong economic implications in other areas besides color communication. The other applications are derived from what is considered the very heart of such a system – the spectral base for color. Contrary to most CAD-type systems, the input and output channels are spectral reflectance values, either measured or generated, and are largely device and illuminant independent. The spectral data are by far the most basic characterization of an object's color. From these spectral values, we derive all the other higher-level output forms such as colorimetric values (X, Y, Z, L*, a*, b*, C*, H*), output to the monitor in calibrated color (R, G, B), and output of the calibrated printer in C, M, Y, K. By combining the spectral base, colorimetric functions, and an image processor, the color imaging system is a powerful tool for color management (Randall 2004).

3. Advancement in Color and Design

Recently, weaving machine producers have achieved a number of technological advancements like high-speed weaving, higher levels of automation, new shedding concepts, automatic (on-the-fly) pattern change, and filling color selection. Along with the advances in weaving, significant development has also occurred in the field of CAD systems, which enables further automation in the design process. Despite this automation, the process of assigning weaves/colors is still done by the designer or the CAD operator, which therefore requires physical sampling prior to production. This section includes the recent research work done to automate the process of assigning weaves/colors in order to reduce or even eliminate the need for physical sampling and to assist woven fabric designers in the creation of pictorial fabrics that are a very close match to the original "artwork" or target.

In woven fabrics, which are highly textured, various patterns become visible through their different structures. The color of such patterns also depends upon the color of the yarns involved, their combinations, and different structures on the pattern surface. The final visible color on the fabric surface is mainly due to the contribution of fabric covering properties, namely optical cover and geometric cover (Lord 1973; Adanur 2001, Peirce 1937). The optical cover properties are defined as the reflection and scattering of the incident light by the fabric surface and are a function of the fiber material and fabric surface. The Geometric cover (characterized by the fabric cover factor) is defined as the area of fabric actually covered by fibers and yarns. The Fabric cover factor is the ratio of surface area actually covered by yarns to the total fabric surface area (shown in Figure 4.19).

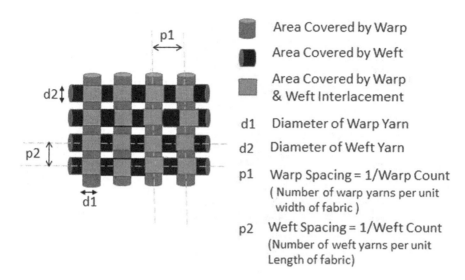

FIGURE 4.19:   Cover factor calculation for a plain weave fabric [10].

By using fabric geometry, fractions of individual color components in a color repeat are calculated and a software tool is then used to calculate color difference tolerance. Experimental fabrics composed from single-colored warp and filling yarns result in weave designs in which warp is interlaced with weft and vice versa [10]. However, a weave design with varying warp/filling colors and diameters will have more than two units, which was not explained in this study. Also, no specific explanation (assumption) regarding yarn diameter and yarn spacing was provided. For calculation purposes, yarn diameter was measured (using a microscope), which actually requires weaving a fabric and hence defeats the purpose of predicting color proportions.

The schematic flow of the design process using the model is illustrated in Figure 4.20. The computer simulation of the model allows the user to enter the design parameters. Next, the developed geometrical model calculates the contribution of each color and, in combination with the color mixing equation, the final color of an area in the pattern can be obtained. The calculated color attributes are compared to those measured from the artwork. The difference of color attributes between the measurement and the calculation is checked. If the difference is within the tolerance, the program reports output that includes the color attributes for calculated and actual, color arrangement, and specific weaves within the classified weaves. If the color differences are out of tolerance, the program reports to the user and suggests possible changes to the input parameters. This iteration continues until a reasonable match for each color in the artwork is achieved.

4. Modernization of Weaving Machines

To keep up with the pace of changing demands in the apparel and fashion industry for cost-effective solutions, there has been advancement in the existing

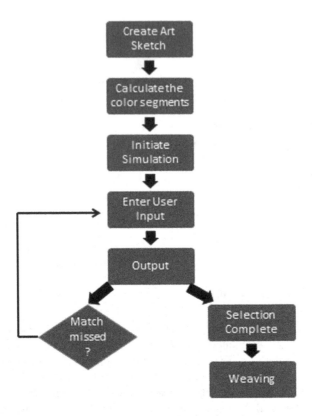

FIGURE 4.20:   Model view of the design scheme of patterned woven fabric.

machinery used for the manufacturing of textiles. From fabric analysis equipment to spinning and weaving to design of multiple textile products, companies are expanding their infrastructure to include automation technologies including but not limited to modernized shuttle-less loom weaving technology that offers multicolor selection with quick time to market. Furthermore, textile industry players are leveraging AI-enabled digital technologies for automatic fault detection and change configuration in no time. Also, using self-optimization methods in the product development lifecycle, the productivity of manufacturing companies and the quality of goods can be increased.

5. **Digital Jacquard Technology**

   Digital jacquard technology is the basis of electronic jacquard machines and new-generation weaving looms, which provide a technological basis for the innovative design of digital jacquard. The design and production processes are both subject to total digital control, and the design data of the jacquard fabric from design to weaving is all processed, controlled, and transmitted in the computer [6]. The technology of developing jacquard fabrics works in accordance with the theory of color mixture in color science, perfectly combining fabric structure and the color of the material. The

FIGURE 4.21: Innovative design processes for digital jacquard.

theory of digital jacquard is applicable to digitizing the objective images with digital technology, and to putting these digital images under innovative digital design directly into structural design. This design concept and method have thoroughly removed the constraint of hand drawing, injecting innovation with jacquard into the overall process of designing; the extent to which the innovative digital jacquard will advance will be extended. The design processes are shown in Figure 4.21.

This study of designing digital jacquard fabric in a digital mode often integrates the basic principles of fabric science, color science, and computer science, in comparison to traditional jacquard fabric, which poses restrictions on the expression of colors. In addition, the scope of innovative jacquard creation is expanded by using a layered combination design method, injecting fresh concepts and know-how into the modeling and coloring of the pattern of jacquard fabric. This provides not only more favorable conditions for the development of innovative digital jacquard fabric but also an effective theoretical basis for the future study of the intelligent design of jacquard fabric with a computer. As the concepts for the innovative design of digital jacquard fabric are being pushed forward for constant improvement, a new design trend is emerging toward innovative digital jacquard design and creation [5].

4.10.4 Design and Technology in Processing and Finishing

4.10.4.1 Designing Technology in Textile Processing

Continuous technological developments are indispensable in textile processing in order to meet the demands of hyper-conscious consumers and to project ourselves as competitors in the global market. Innovative technologies such as electrochemical technology, plasma technology, Nanotechnology, and chemical technology are anticipated for a better prospectus with due regard for their ecofriendly nature, energy efficiency, and best quality performance.

1. Textile Dyeing:
   Dyeing is thriving scientifically as it proceeds to walk along the automation path; every single minute parameter is very critical. Even milligram-level recipe variations can change the shade of the fabric. All the parameters for dyeing – temperature, pressure, water level, water flow, circulation, and time of treatment – are very important. Automation of the dyeing process can improve productivity by controlling these parameters very accurately. Incorporation of science by bringing automation into textile dyeing and printing means one or more (or all) of the following steps:

- Programmable process control (by microprocessors) of the machinery;
- Dissolving and dispensing of the dyes, pigments, and chemicals in a central color kitchen;
- Computer-controlled weighing of solid material with automatic stock control and the printing of recipe and process cards;
- Color measurement, computerized color matching;
- Central computer (network), computerized management system

2. Textile Printing:
   Textile printing is the process of applying color in definite patterns and designs. Fabrics that are properly printed have color bonded with the fiber, so as to resist color fading while being washed or having friction applied. Hand block printing, tie and dye, digital printing, and screen printing are the common ways by which these prints can be incorporated on the fabric to make suitable patterns like stripes, paisleys, polka dots, and abstract prints. These prints add a striking feature to the garment and make it look alluring.

4.10.4.2 Designing Technology in Textile Finishing

Automation in the textile finishing industry is not a new concept, but it is being modernized day by day. The textile factory is characterized by a considerable fragmentation of the production cycle into a number of segments specialized in the production processing of different fibers/yarns. The objective of modern automation technologies is to achieve flexibility and quality by the following three reliable paths:

1. The automated standardization of components;
2. The automated compatibility of systems;
3. The popularity of personal computers in the case of textile finishing.

In the textile industry, finishing is usually done in the final stage of textile processing, as a result of which the textiles gain several functional characteristics. A wide variety of finishing chemicals are now available in the market that meet or exceed the expectations of the consumers. Novel finishes providing high-value additions to apparel fabrics are greatly appreciated by a more demanding consumer market. The various textile-finishing chemicals in Table 4.1 are used to convert a textile material into a technical textile with impeccable functional properties.

4.11 FUTURE OF TEXTILE SCIENCE

A wide range of textile products is available in the market due to the advent of science and technology and less dependency on the weather. The increasing quality of textile products is an indicator of a modern world which gauges the selling prospects of a product in the market. Nevertheless, the textile industry has always aced this measure by occupying an important rank in its contribution to the gross national product, employment, and export revenue. Today, the textile industry is serving a key role in lifting the economy and technological advancements by making room for humanized design and transforming production methods toward achieving human-sensitive textile engineering.

TABLE 4.1
Types of Finishing in Textiles

Finishing Technique	Finishing Agents	Mechanism	Application
Thermal regulation	Microcapsules/phase change materials/chemicals such as nonadecane and other medium-chain-length alkanes	Maintain core body temperature by storing or releasing latent heat energy to change its phase and provide comfort to the wearer in diverse environments	Aerospace, automotive, agriculture, biomedical, defense, sports, and casual fabrics
Easy care	Compounds of low formaldehyde resin, polyethylene, and silicones	Involves release of low formaldehyde; chemicals are applied through a cross-linking effect for prevention of wrinkles	Maintain image of cellulosic fabrics by improving crease recovery in high-performance textiles
Flame retardant	Wide range of flame-retardant chemicals based on the application	Chemicals added to flammable materials to increase their thermal protective performance and to render excellent durability to multiple dry cleans	High-performance fabrics blended with Nomex, Flame-retardant rayon/cotton, and high-grade Basofill
Superabsorbent	Chitosan derivatives grafted with acrylic monomers have promising applications in various fields as superabsorbent	It acts like a super-sponge and can absorb aqueous fluid many times its own weight, forming a gel which does not subsequently release it	Baby diapers, incontinence products, etc.; useful in agriculture, biomedicine, sanitation, geotextiles, and protective clothing
Medical, cosmetic, and odor-resistant	**Medical** – bioactive compounds; **Cosmetic** – vitamins; **Odor resistant** – cyclodextrins	All three finishes are released through active compounds and transferred to the wearer's skin in a controllable manner	Widely used for medical, hygienic, health, and cosmetic purposes
Hydrophobic and oleophobic	Polysiloxane, Fluorocarbon compounds, and nanomaterials	Involves conventional padding and exhaustion; Plasma techniques are used to develop super-hydrophobic substrate	Development of water-repellant and oil-repellant textiles, providing lotus effect to surfaces
Ultraviolet and radiation protection	Fine particle of carbon black, phenylbenzotriazoles, UV531/TiO_2, benzophenones	Fabrics are treated with UV inhibitor agents that absorb energy in ultraviolet range of electromagnetic spectrum, before the radiation is absorbed by polymers	Improving radiation-protective properties of fabrics, e.g., sun-protective factor; preventing degradation of fabrics by harmful radiation

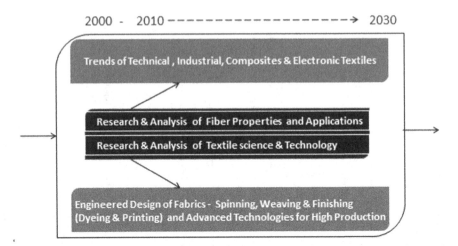

FIGURE 4.22: Future prospects for advancement in textiles and manufacturing technology [9].

The framework of textile technology and apparel design is forecast to shift toward engineered design and high-performance manufacturing technology by 2030, with ongoing research on fiber and textile mechanics. To realize this transformation, the collaboration of academicians and industry professionals is paramount in broadening the scope of this fast-expanding realm and in creating a new culture of innovation. Meanwhile, the continuous efforts by clothing science to tackle challenges related to health and safety, environment, and improving the quality of people's lives will continue to lead to new technologies and the discovery of smart materials.

Through the illustration in Figure 4.22, it is clear that textile reconstruction at advanced levels will reap benefits for researchers and textile technologists when the design and selection of raw materials are systematically organized. The future of the garment industry is largely dependent on engineered manufacturing processes to produce high-quality textile materials. To accommodate sustainable technology, designers and clothing technologists need to focus on the principles of circular economy to save on fiber and fabric wastage. The industry should focus on high-quality fabrics of reasonable price instead of producing poor quality.

Undoubtedly, the use of fibrous composite materials and technical textiles is the future of the next-generation textile industry. These two terms are echoing throughout the world of fashion and textile, with their engineering applications and properties based on the advanced theories of textile mechanics.

SUMMARY

From the Stone Age to the Modern Age, Science in textiles has played an integral part in the growth of human society and has had a profound impact in shaping our lives. Textiles are being increasingly consumed as lifestyle products, with consumers becoming more ergonomically conscious and value focused than price sensitive.

With unprecedented levels of growth and investments taking place with evident limited natural resources, the clothing industry needs a breakthrough science innovation to create a uniform system of production, consumption, and dispersal of materials. This certainly will push fundamental research on material science, fiber, and fabric mechanics toward more advanced levels [9].

In terms of the role of textile research and clothing science, it is imperative that the field should not become stagnant if it is to be successful. In the midst of the rapid changes that occur in our high-information society, textile design needs to have strong connections between scientific research and technology for the production of fabric by techniques like spinning, weaving, and dyeing/finishing; this would certainly become the power to sustain the industry for a long time. The invention and development of textile fibers combined with sophisticated electronics and computerized systems have transformed and revolutionized possibilities for distinct smart textile materials capable of responding to the environment. There have been so many advances in the last few years; it is difficult to predict what the future holds, but one thing certain is that textiles will continue to influence our lives for the future of humanity.

REFERENCES

1. Cormac McGarrigle, Ian Rodgers, Alistair McIlhagger, Eileen Harkin-Jones, Ian Major, Declan Devine and Edward Arche. "Extruded Monofilament and Multifilament Thermoplastic Stitching Yarns", *Fibers* (2017), 5(4): 45.
2. https://www.macmillanhighered.com/BrainHoney/Resource/6716/digital_first_content/trunk/test/hillis2e/hillis2e_ch04_5.html
3. C. Atkinson *False Twist Textured Yarns: Principles, Processing and Applications*, Cambridge: Elsevier, (2012).
4. Zhigang Xia and Weilin Xu. "A Review of Ring Staple Yarn Spinning Method Development and Its Trend Prediction", *Journal of Natural Fibers* (2013), Taylor and Francis 10(1): 62–81.
5. M.C.F. Ng and J. Zhou *"Principles and Methods of Digital Jacquard Textile Design", Innovative Jacquard Textile Design Using Digital Technologies, Woodhead Publishing Series in Textiles*, Cambridge: Elsevier, 22–48, (2013).
6. A.R. Horrocks and S.C. Anand. *Handbook of Technical Textiles*, Cambridge: Elsevier, (2000).
7. M. Tausif, T. Cassidy and I. Butcher. "3-Yarn and Thread Manufacturing Methods for High-Performance Apparel, High-Performance Apparel Materials, Development, and Applications", *Woodhead Publishing Series in Textiles* (2018): 33–73. https://www.sciencedirect.com/science/article/pii/B9780081009048000031
8. https://www.arahne.si/products/arahpaint/
9. Masako Niwa. "The Importance of Clothing Science and Prospects for the Future", *International Journal of Clothing Science and Technology* (2002), Emerald Group Publishing Limited 14(3l4): 238–246.
10. Kavita Mathur and Abdel-Fattah Seyam. "Color and Weave Relationship in Woven Fabrics", *Advances in Modern Woven Fabrics Technology, Intech* (2011). https://www.researchgate.net/publication/221913744_Color_and_Weave_Relationship_in_Woven_Fabrics

5 Innovative Graphics – Enabling Change by Thinking Science

5.1 INTRODUCTION

Innovation is what drives the digital world ahead by forging technological and cultural advancements in order to address simple and complex practical problems. The beneficiaries of this innovation are the individuals who are looking for quick services and more efficient products. On the flip side, these beneficiaries may constitute an organizational framework looking for competitive success in the market.

By nature, innovation necessitates a response to a particular problem domain in context. In other words, driving innovation always requires a medium, which may be in the form of imagination, a requirement, or a catalyst that fosters creativity for improvement in the system. Design is regarded as one of the basic catalysts of all forms. Establishing a relationship between innovation, design, and creativity, innovative graphic designing has paved the way to transforming the communication of ideas through the use of digital tools and capabilities. Involving the flavor of science and research in the thinking of innovative design can produce profitable and sustainable benefits to businesses and individuals beyond their perceptions.

5.2 UNDERSTANDING GRAPHIC DESIGNING

Design does not happen solely on a computer to create a product. There are always background theories and logical thinking that make design as a project possible. These theories and logic are built on scientific studies that correlate with the needs and benefits of a designing project. The connection of design and science simplifying complex and often abstract ideas to design methods used by scientists to get their point across is a tale as old as time. Over the last two decades, design research has gained momentum and respect within the academic community, but unfortunately, research in graphic design has not been regarded as a significant player. Often people take research on graphic design as something that is supplementary or additional to other projects. This section explains about the more positive view of graphic design research and the need for graphic designers who work with teams of illustrators, photographers, coders, editors, and writers by using graphics language in verbal, pictorial, and schematic form.

There is a contrast between the concepts of graphic designing and graphic design research. *Graphic Designing* is a planning and problem-solving discipline, concerned with purpose, visual judgment, process, engagement with users, and circumstances of use. On the other hand, *graphic design research* is a form of professional practice that is concerned with overall quality more than with an individual work produced by a graphic designer. There are certain areas of graphic design research like typography, way-finding, periodical design, interaction design, illustration, exhibition, and branding of corporate identity which not only enhance the aesthetics but also imbue functional properties into design. Within each of these areas the human-centered approach becomes imperative in studying the role of people, locations, history, cultural theories, and practices in graphic design. *Graphic design research is the key to providing different kinds of evidence of the way design decisions are made, and of the way graphic software is used.* The framework of graphic design research for large projects in service design, information design, and user experience design encompasses the role of innovative graphic design in paper and digital form, in order to scale interaction with the environment, elucidating the process for the better understanding of design and meeting project requirements by the designers.

5.2.1 Elements of Graphic Design

In the quest to turn innovative ideas into real graphics, designers should always fasten their seat belts to take a tour of endless possibilities and to tap into the untapped space of graphic creations. Each possibility holds a fundamental sphere of elements that are instrumental and must be comprehended for enabling innovation in the graphic design process. These elements – Innovation, Creativity, and Design – are illustrated in Figure 5.1 and are described in detail forthwith.

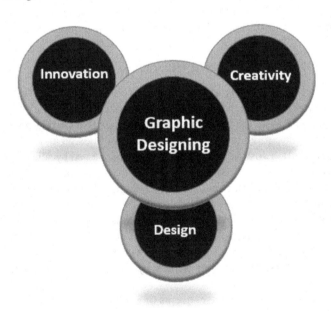

FIGURE 5.1:   Elements of graphic design.

5.2.1.1 Innovation

Many innovative design experts believe that innovation is a phenomenon that can be transformed from a random event into a series of scalable tasks. The core of innovation in a graphic design lies in simplifying the problem and recognizing what a good idea should look like. In the modern technology era, the focus of innovation has shifted from being product-centric to customer-centric, from engineering-driven to design-driven, and from marketing-focused to user-experience-focused. This is today achieved using a new tool – design thinking – which is at the core of effective strategy development and organizational change. In other words, design thinking allows action-oriented and solution-focused designing, rather than problem-focused designing.

Design Thinking – A Tool for Innovation

In the midst of uncertainty, we need a tool for continuous innovation in the domain of graphic design; this tool is referred to as design thinking. It is a user-centered approach of solving difficult problems to create desirable innovative products that are profitable and sustainable over their lifecycles. Design is an iterative process, and design thinking is present in each stage of the journey from client brief to finished work. Over recent decades, it has not only been designers that have been using design thinking as a tool to understand and act on rapid changes in our environment and behavior; managers are also trying their luck with this magic wand to address ill-defined or unknown problems. Thinking scientifically helps designers to carry out the right kind of research, create prototypes, and test out products and services to uncover new ways to meet users' needs.

Design thinking methods and strategies belong at every level of the design process. However, design thinking is not an exclusive property of designers—*all* great innovators in literature, art, music, science, engineering, and business have practiced it. What's special about design thinking is that designers and their work processes can help us systematically extract, teach, learn, and apply these human-centered techniques in solving problems in a creative and innovative way: in our designs, in our businesses, in our countries, and in our lives.

That means that design thinking is not only for designers but also for creative employees, freelancers, and business leaders. It's for anyone who seeks to infuse an approach to innovation that is powerful, effective, and broadly accessible, one that can be integrated into every level of an organization, product, or service so as to drive new alternatives for businesses and society. While the complete process of design thinking boils down to the following five steps which are not sequential, designers can run the stages in parallel as per the demand of the time, or execute them out of order and repeat them in an iterative fashion (Figure 5.2).

Stage 1. Empathize – Research Your Users' Needs

The foremost stage of the design thinking process is to empathize with its customers by standing in their shoes or seeing through their eyes. This gives an opportunity to find the solution for an untapped part of the problem. It can be done by conducting surveys and/or face-to-face interviews to learn the views of other people solving the same or similar issues in the context of the problem. There are a number of

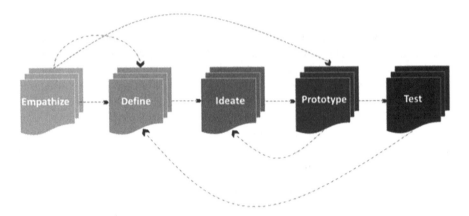

FIGURE 5.2: Stages of design thinking.

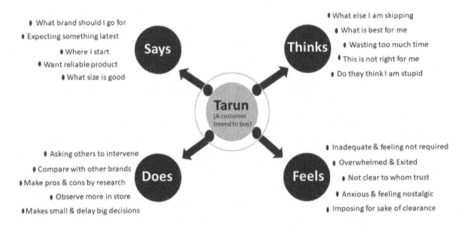

FIGURE 5.3: Illustration of an empathy map.

tools available to consolidate valuable information directly from the users. Creating empathy maps is one of the coolest tools being used to capture reactions from different people to learn about what people *do*, *say*, *think*, and *feel* in the context of the problem. Figure 5.3 indicates the tools of the empathy map, which not only provides a particular direction in identifying a problem but also leads to finding an ideal solution.

Empathy maps provide more than just a way to summarize data. They also help colleagues understand the context of the problem and how people experience it, too. Arming everyone with this knowledge helps ensure that the person the solution is intended for is top in mind at all stages of the idea's development.

Stage 2. Define – State Your Users' Needs and Problems

Collecting the inputs from the user research leads to the next important stage of design thinking, i.e., defining a problem statement. A problem statement means a concise description of a challenge or a problem to be addressed or a situation to be improved upon. Defining a problem statement allows recognition of the gap between the desired state and the current state of a design process or a product. Focusing on the facts, the problem statement should be designed to capture human-centered terms rather than to focus solely on business goals. A problem statement defined on the basis of empathy proves beneficial for building real relationships with the customers and enhancing their experience with the business products. One such strategy is illustrated as follows.

Example: A parenting e-commerce business wants to increase its customer reach and design a product model that addresses the objective of revenue generation and customer retention.

Solution: Using empathetic research to capture the real situation of the customer, and then defining a problem statement instead of directly setting a design goal based on business requirements only, will create a win-win situation for the business in both ways.

One of the approaches includes a human-centered target to provide working mothers or parents with prototypes of health products to keep their families fit. In this scenario, the beneficiaries will be led to create an account on business websites, as they will be looking for more ways to balance their health, work, and life. This strategy will absolutely work better for a business to improve the product design based on the user's feedback and new ideas.

Last but not least, continuous research is the key to success in problem formulation, which brings designers to a quick start on Stage 3.

Stage 3. Ideate – Challenge Assumptions and Create Ideas

Now that the problem is apparent, it's time to brainstorm ways to address those unmet needs. The ideation stage marks the transition from identifying problems to exploring solutions. The solid background knowledge accumulated from the first two phases allows you to "think outside the box", look for alternative ways to view the problem, and identify innovative solutions to the problem statement you've created. Here we prioritize breadth over depth as we look for a diverse range of ideas to prototype and test with real people in the following two steps.

When ideating, challenge yourself to go beyond minor adjustments. Prototypes provide a way to investigate riskier ideas cost-effectively, and the testing phase provides more confidence that the risk is worth pursuing.

The ideation stage flows between idea generation and evaluation, but it's important that each process remains separate from each other. On a reality note, when it happens to generate ideas, we must draw a sketch or make a note without bothering with the quality of sketch or handwriting, because that rough idea itself can turn into a successful business. After the collection of all sorts of ideas, they need to be filtered down for further clarification in order to meet the desired objectives. The top ideas with the maximum number of voted dots move to the next step of prototyping.

Stage 4. Prototype – Start to Create Solutions

Prototyping allows the designers to reach closer to their product by converting their ideas into physical form and to gain feedback from the people they are intended to serve. The aim is to start with a low-resource version just to check the feasibility of the product for the target solution and get it improved before the deadline based on each and every user feedback. Beginning with a paper blueprint can help you learn quickly with minimal effort. At the stage of prototyping, it's often a good idea to work through the assessment of the prototype in depth to ensure that any significant gaps are identified before the prototype is delivered for user acceptance testing in the next run.

In context, the prototype should be a realistic representation of the solution that allows designers to gain an understanding of what works and doesn't work. It also undergoes changes and updates on feedback from the testing phase in multiple iterations. A low-cost, simplified, strategic, and fine prototyping allows the team to develop and test different versions to reach the right solution that meets unrealized user expectations.

Stage 5. Test – Try Your Solutions Out

Once you get off the ground with the testing phase, the core and crucial activity of design thinking commences. Testing a solution prototype commemorates the conclusion phase of the product development lifecycle. After qualifying standard user guidelines, a prototype in the testing phase acts as a conversation starter to gain deep understanding of the pain points, and to determine whether the proposed solution is suitable for the problem in context.

The Stanford school of design thinking encourages practitioners to **"Prototype as if you know you are right, but test as if you know you are wrong".** Testing happens in two forms: validation and verification. Inspection of design capabilities as per the requirements by means of end user feedback validates status on whether or not it solves their problem. On the other hand, testing of the operational performance and technical efficiency refers to verification testing.

Since design thinking is a customer-centric process, versions of the prototype are iterated upon and then reintroduced to end users fetching more feedback to improve the solution. Openness to challenges and feedback is needful even if it requires a start from scratch. Thus, design thinkers in action should be prepared to build the new version if the prototyped solution does not adequately address the problem. The testing phase is enacted to deduce whether the issue was framed wrongly from the beginning.

The complete cycle of the design thinking process is executed with a target objective to do the upfront work to justify a solution that addresses a real practical problem while gaining deep understanding of the behavior and expectations of the people who might use it, so there's a higher likelihood of a successful product or customer service in the long run [1].

5.2.1.2 Design

A significant shift in the range of approaches to contemporary design has come about as a result of the debate surrounding the notion of graphic authorship and what this might constitute in the future for graphic designers. While definitions of authorship in graphic design continue to be expanded and updated by designers, design writers and educators, it is useful to consider a singular interpretation as a starting point for debate (such as a train timetable or road sign) or to persuade a target audience about a particular product through its packaging and promotional design. While this may be a crude definition, it is clearly applicable to the broad majority of design practices in the commercial arena: graphic designers are commissioned to employ their skills as communicators in the service of a client.

The notion of authorship lies in the possibility that designers can also operate as mediators – that they can take responsibility for the content and context of a message as well as the more traditional means of communication. The focus for the designer might be on the transmission of their own ideas and messages, without the need for a client or commissioner, but sill remaining fixed on the effectiveness of communicating with an audience. This can also operate in a commercial sense – a client might choose to employ a graphic designer because they have a particular visual style or method of working that would work in tandem with their message or product. This could be described a designer's signature style, and there are many celebrated or well-known designers who are commissioned purely because of a body of work that is concerned with particular themes or is popular with a particular audience.

Traditionally, graphic designers are involved in a process of facilitation: put concisely, the business of design is to communicate other peoples' messages to specified audiences. This might be for the purposes of providing general information.

Science and Design Research

Research methods can be defined as ways of approaching design problems or investigating contexts within which to work. This chapter focuses on the thematic approaches to problem solving and the construction of rational and logical systems of design thinking. Through knowledge of existing conventions and the development and application of a personal visual vocabulary, designers are able to make more effective use of their perceptions and discoveries, and to work practically and creatively with reference to a wider cultural context. Systematic research methods encourage designers to develop a personal and critical point of view through the recording, documenting, and evaluating of visual and verbal structures, languages, and identities in the wider environment, and then applying those findings within their own work. The adoption of a rigorous methodology that either addresses the specific requirements of the brief or sets a series of boundaries within which to work on a broader visual investigation can help the designer to focus a project and define the exact problem, or series of problems, to address. Breaking the project down into a set of intentions, each with defined parameters and a predetermined level of background knowledge or experience on the part of the designer, makes the task more achievable and the goals of each stage of the process more explicit.

Each of these areas will be explained in detail within this chapter, showing the developmental process of a strategic design methodology relevant to the context of the brief. Examples of work illustrating key concepts from both the professional and academic fields are included to guide you through each stage of the process, and to help define each specific area of investigation undertaken by the designer. The first task for the designer is to identify what he or she is attempting to achieve with the project. Within commercial graphic design, this might be described in the brief as the intended message that is required to be communicated, or the target market which a commercial enterprise wishes to engage with. In this instance, the work undertaken is a form of applied research. Any design brief can be broken down into three areas for specific interrogation: *a field of study, a project focus,* and *a research methodology.*

Field of Study

This area describes the broad context for the work. This could be, for instance, the field of way-finding and signage within information design, or an audience-specific magazine page layout aimed at a particular cultural group. The first task facing the designer is to research their field of study, to acquire knowledge of what already exists in that area, and to determine the visual languages which can be directly associated with the specific target audience or market for the design. The designer needs to consider both the external position of their intended work (the explicit aim of the communication itself) and its internal position (the relationship between this particular piece of visual communication and others within the same context).

This is very important, as contemporary cultures are saturated with displays of visual communication, in the form of advertising, information graphics, site-specific visual identities, and images related to entertainment or decoration. If a piece of graphic communication is to be displayed within this arena, the designer needs to be aware of how it relates to competing messages, and how the problems of image saturation or information overload might be resolved in order to communicate effectively.

Of course, the designer will become more familiar with a specific field of study through professional experience. By building a relationship with a particular client and working on a number of projects aimed at a distinct audience or cultural group, the designer can learn which forms of communication are likely to be more (or less) effective. Field of study research then becomes more intuitive, based on prior experience and learning, and the designer is able to move more quickly toward a project focus and methodology.

Field of study research takes a variety of forms, dependent on the context of the work. Market research might be appropriate to some briefs, whereby the designer seeks out other work in the same field and analyzes and compares the visual forms of communication relevant and readable to a specific audience. This could mean a review of comparative products or visual systems, working with a client to establish their position in the marketplace or their aspirations to communicate with a particular audience. In most cases, sophisticated visual languages already exist which attempt to engage those audiences, and the designer should become familiar with their vocabulary, even if his or her intention is to create a new form of communication which sets itself in opposition to that which already exists. Cost implications are also important to consider at this stage of the project. The costs of materials, print reproduction or other media (web design, digital storage, etc.), labor, and overheads need to be taken into account against the intended budget for the project. The designer and client need to have a strong idea of the range of materials available and affordable to them, and the implications upon the design itself. If the budget can only cover the cost of two-color printing, for instance, then those restrictions need to be put in place in advance and then turned to the designer's advantage in seeking innovative responses to the range of techniques and materials available.

Project Focus

Once the designer has researched the field of study and become familiar with the broader intentions of the brief, a specific project focus is necessary in order to demarcate the exact intentions of the work to be undertaken. At this stage, the designer should be able to describe the message which is to be communicated to a specific audience, or within a specific context, and the aims and objectives of that communication – for instance, to persuade the receivers of the message to act in a particular way (e.g., buy this product, go to this event, turn left at the next junction), or to clearly communicate a particular emotion or identify with a subcultural group.

The focus may change during the lifespan of a project, becoming broader and then being redefined in an ongoing process of critical reflection and reappraisal. There are a number of ways in which the narrowing down and refining of a project focus might take place. Two useful models for the designer to use in order to ascertain the context of their work and define a particular focus are set out diagrammatically in the following pages.

The first research model, which we will term the "Context-Definition" model, gives a greater emphasis to the investigation of a field of study. In this model, the designer attempts to become more expert within the field of the brief and the project focus is defined in response to an identifiable need within that area. The second model, termed "Context-Experiment", still requires the designer to undertake a broad preliminary analysis of the field of study, but the practical work on the brief itself begins earlier in the process than in the "Context-Definition" model. Usually, this is done through a series of tests or experiments which can be evaluated within the field of study, leading to a redefinition of the project focus dependent on the results gathered.

It is important here not to lose sight of the original project intentions, and to work through the experiments in a systematic way. The "Context-Experiment" model will inevitably lead to a number of "failed" experimental outcomes, as each small "test" is an attempt to gain feedback in the definition of the project focus. In fact, if an experimental piece of visual communication is unsuccessful when tested with a target audience or in a specific context, then this should still be seen as a positive exercise in gathering information on the project focus. By determining what does not work, as well as what potentially does, the designer is in a far better position to arrive at a more successful resolution.

Research Methodology

A research methodology is simply a set of self-imposed rules by which the designer will engage with a project. Once the intention of the work has been clearly stated, together with a detailed mapping of the field of study and the definition of a focus for the visual message to be created, the designer needs to outline exactly how he or she intends to go about developing the project and testing ideas in order to create an effective solution to the brief. The intention here is to develop systematic ways of working which lead progressively to a more successful outcome, based on experiments and visual testing, materials investigation, and audience feedback, and the goal is to produce a piece of graphic design which is effective, useful, and/or engaging. The adoption of a strong research methodology should help the designer to make work which can be justified in terms of the processes used, and can be predicted to get closer to this goal. It is also important to plan the work in advance, including a rough schedule identifying when the designer expects to undertake each experiment, and the proposed deadline for finishing the project. Whether within the areas of commercial graphic design or design study, deadlines are usually given as a part of the brief. Even where this isn't the case, for instance when a designer is conducting a personal visual investigation, it is still important to plan a time frame for the project.

"Experimentation" is something of a buzzword in contemporary graphic design. An experiment is a test or investigation, planned to provide evidence for or against a hypothesis – an assumption which is put forward in order to be verified or modified. When a designer is working toward producing a piece of work, a series of visual tests or design experiments might be useful in gathering feedback on new ideas and forms of communication. However, experimentation is not a virtue in itself – it has to operate within a set of precise guidelines, delineating the intention and context of the experiment, together with the ways in which feedback will be gathered and results will be measured. In short, a design hypothesis might be that the creation of a particular visual form will communicate a particular message to a given audience. An experiment to test this hypothesis would then involve creating variations of that form and gathering feedback from target audiences or experts within that particular field of design in order to measure the relative success or failure of the work to communicate as intended.

In setting up a series of "experiments", which might involve trial runs with alternative visual strategies in response to a defined problem, how is the designer to go about gathering feedback in order to evaluate which of the visual applications is the more successful?

There are a number of ways to respond to these questions. Market research, especially in relation to product advertising and marketing, has developed some successful methods of testing materials and form through the use of focus groups, statistical analysis of surveys, and audience observation techniques. Some of these techniques can be linked to anthropology and the study of human interaction within social groups, while others derive from more scientific methods of data gathering and quantitative analysis. It is also important to understand the differences between what are termed quantitative and qualitative methods. The designer will often use both forms of analysis, and their application may prove more or less useful depending on the specific brief and target audience, but the methods themselves are distinctly different.

Quantitative analysis is based upon mathematical principles, in particular statistical methods of surveying and interrogating data. By producing a number of successful methods of testing materials and forms through the use of focus groups, statistical analysis of surveys, and audience observation techniques. Some of these techniques can be linked to anthropology and the study of human interaction within social groups, while others derive from more scientific methods of data gathering and quantitative analysis. It is also important to understand the differences between what are termed quantitative and qualitative methods. Designers often use both forms of analysis, and their application may prove more or less useful depending on the specific brief and target audience, but the methods themselves are distinctly different.

Quantitative analysis is based upon mathematical principles, in particular statistical methods of surveying and interrogating data. By producing a number of visual forms to test, the designer can place these objects in specific locations in order to "count" positive and negative responses from a target audience. This could mean conducting a survey within the target audience group, using multiple choice questions devised to score against a set of criteria. The data gathered can then be converted into numbers, to be analyzed statistically to find the most successful visual form. Of course, as the size of survey group or sample increases, we anticipate that the results will become more accurate.

Quantitative methods find a significant role in the areas of data investigation and technology. If a piece of work is to be produced in multiple numbers (as almost all graphic design is), then the criteria for choice of materials – its resistance to age deterioration or discoloration, distortion, stability, and its fitness for purpose – can be subject to quantitative evaluation. Materials testing, through the use of alternative substrates and surfaces on which to print, or technologies with which to construct and view online data, is an important area of design experimentation, and experiments within this area can usually be measured with reasonable accuracy.

Similarly, the cost implications involved in the selection of materials and production methods can be compared and measured against the constraints of the project budget. When a piece of work is to be manufactured as a long production run, especially in printed form, the costs involved in even the smallest design decision are magnified accordingly – from the cost of ink and paper stock to the labor involved in folding, collating, cutting, and finishing.

Printers set up their machines to operate using the most common formats and production runs. This usually implies a reliance on standard ISO (International Organization for Standardization) or imperial paper sizes and color palettes (CMYK four-color process or Pantone spot colors for instance). If the designer chooses to work outside of these standards, set-up times for production will be longer and the costs will, therefore, increase. Any tasks which require hand rather than machine finishing will also incur additional costs. As such, the economic aspects of the project need to be planned carefully in advance, and quantitative methods can be useful for the designer in calculating the budget for the project.

Qualitative analysis in design, on the other hand, is based on subjective responses to visual forms and the reading of graphic material by a viewer. Often this is done by the designer herself, in the form of critical self-reflection. It is also implicit, of course, in the surveys and focus groups mentioned earlier in this section (see above on quantitative analysis). The reading of images and visual signs is a qualitative act in itself; although some responses can be evaluated statistically as a form of quantitative analysis, the initial data gathered is based on human reaction to the visual forms and experiments presented.

A key qualitative method for designers involves the semiotic analysis, or deconstruction, of design artifacts. What this means in practice is the reading of explicit and implicit messages within a visual form, to determine the range of meanings which might be communicated to an audience. If the principles of visual communication are broken down into the twin themes of the encoding and decoding of meaning (synonymous with the acts of writing and reading), then the range of implied messages and interpretations can be largely determined in advance. Graphic design usually operates within very specific boundaries, where the intention of the brief is made clear by the client or designer in advance. Certain vocabularies drawn from communication and language theories can also help the designer to describe the range of activities involved in the process of visual communication.

These methods are useful for the graphic designer, as they can help to build constraints into the visual message in order to guide the viewer toward the desired reading, rather than a misinterpretation, of the message [2].

5.2.1.3 Creativity

Creativity is the skill of connecting hidden dots and patterns for transforming new and imaginative ideas into reality. Creative design needs to be innovative and imaginative, but it must always be rooted in practicality. However, it is not essential for graphic designers to be creative but to create unique methods to address complex problems with the mindset of a designer. Hence, creativity is one of the prominent elements of graphic designing in satiating the needs of the consumer by ensuring the usefulness and market acceptance of ideas and products. For example, the creativity in smartphone communication has evolved from big devices to compact devices that we can carry in our pockets or even wear around our wrists.

Creativity in graphic design is described by its art, nature, originality, and critical thinking.

- *Creativity by Nature*: If the act of designing is a creative act, then anything that goes into that act is a creative process. By its nature, graphic design is broadly creative. This may be a too-convenient definition, however, as every process in a creative act can't be considered creative, particularly, if it is used repeatedly in a process-like flow.
- *Creativity by Originality*: Creative geniuses produce completely original ideas. This definition of creativity is not the process of creation, but the power of designers to conjure original thoughts and implement them in novel ways. It can also mean expanding on the established rules to create something wholly different from its original form.
- *Creativity by Critical Thinking*: Critical thinking provides the filter that sifts through the creative process and pulls out what is usable and what isn't. Graphic designers should be ready to use both their creativity and important thinking skills together. This can be in two stages, where the creative part is an "idea dump" of sorts and the critical thinking is the sifting through to find what sticks. Or it is often done simultaneously during a flow-like state of concentration that true designers master.

- *Creativity by Art*: The graphic designer should have an artistic sensibility. This sensibility is what drives the designer to use certain colors, compose design art in certain ways, and use specific sorts of typography. How designers use their artistic sensibility depends on how creative they are in wielding it. Unlike critical thinking, which filters creativity, artistic sensibility multiplies it.

5.3 FOUNDATION OF GRAPHIC DESIGNING: INNOVATIVE THINKING

Design is the general term for execution of ideas and plans. According to the design space situation, it can be classified into graphic design, three-dimensional design, and space design, of which graphic design is the basic one. Combining art with science gives graphic design aesthetic character and practical applicability. Generally speaking, the essence of graphic design is aesthetic creation with innovative implementation as its foundation. Originality that guides the aesthetic creation implementation has a fundamental role in graphic design. Originality refers to designers' consciousness of innovation, which is achieved by their innovative thinking. So as we said, innovative thinking is the heart of originality and the essence of graphic design.

5.3.1 RELATIONSHIP BETWEEN INNOVATIVE THINKING AND GRAPHIC DESIGN

Innovative thinking is at the heart of graphic design and three-dimensional design. Design usually means a purposeful and thoughtful plan, which would be shown by graphic design, and is an artistic form that hews closely to a specific purpose. In graphic design, we generally use a visual element to express the designer's ideas and plans and make use of text and graphics to transfer the information to customers in order to allow people to understand our vision and plan. No innovative thinking means there is no connotation. That is to say, the whole process of design activity is a process of formation of design work through design and conception from innovative thinking. Innovation focuses on discovery and creation, which is an essential attribute of graphic design. The significance of innovation lies in breaking current constraints to produce design conceptions through the originality and novelty of the ideas or forms as well as to produce design thinking by handling design processes, accumulating perceptual data, and applying corresponding thoughtful and imaginative means to achieve rules processing. For design selection, the formation of innovative thinking will improve designers' capabilities. And yet accurate and scientific innovative thinking can promote the development and progress of graphic design.

5.3.2 ROLE OF INNOVATIVE THINKING IN GRAPHIC DESIGN

5.3.2.1 Improve Innovation Capability

As one of the important aesthetic creation activities, the quality of graphic design is determined by designers' innovation capability. One should improve innovation capability through the development of innovative thinking to increase design effect.

As one of mankind's unique aesthetic creative activities, graphic design requires designers to employ innovative thinking methods. In line with other art design, its essence lies in aesthetic creation. Only designers pay attention to training their innovative thinking, and improving their aesthetic innovation capability can help them adapt to the aesthetic creation needs of this era. Creativity is an ability that people generally have. However, designers should focus on how to improve their innovation. According to one study, enhancing designers' innovation could be realized by training their innovative thinking ability. Yet the implementation of the initiative of creation can become the top priority of artistic creation. Initiative is usually generated from training designers' innovative thinking ability during practical activities rather than being spontaneously formed.

5.3.2.2   Enhance Originality and Vitality of Graphic Work

Innovative thinking is an important characteristic in graphic design. First, as a practical activity limited by time and space, major works of graphic design will inevitably be marked with the characteristics of a particular era. The designers will enhance modern consciousness and innovation consciousness to break with traditional thinking. Only being in an open pattern, training innovative thinking, and breaking limitations of time and space can realize breakthroughs in art, which could reduce the negative impact of limitations of time reasonably, thereby enhancing the vitality of design works. Second, graphic design is an innovation activity that meets certain requirements of modern people, which shall be implemented according to their modern aesthetic. People also have different requirements, which makes designers use their innovative thinking to satisfy this diversity of aesthetic demand. Third, the essence of graphic design is innovation. Graphic design activity itself is a kind of aesthetic creation, which requires designers to have the spirit of innovation and the ability to make great efforts to achieve.

5.3.3   IMPACT OF INNOVATIVE THINKING ON GRAPHIC DESIGNING

The depth of design thinking becomes a key success factor in graphic design works. In the course of teaching, we found that many works had a strong sense of form but weak innovation. This obvious phenomenon in professional design may be caused by ignoring the train of innovative thinking. Hence, designers should consider several issues like how to train and apply innovative thinking, how to make works with contemporary and aesthetic feeling, and how to made works with meaningful form. For understanding the objective world, modern people usually have two different ways of thinking, that is, logical thinking and image thinking. The former is a progressive linear thinking process, which uses abstract conceptualizations of visualized content to form concepts or theorems and to realize intended purposes through abstract inference and judgment. The latter mainly has specific images and representations of objects as its main content. It uses the accumulation of specific images and abandons nonessential sensibility; then, it uses imagination to deduce the typical image. It mainly uses images as materials and tools to create a character. Based on these means, it has nonlinear characteristics with non-continuity and jump.

The main implication of design thinking is innovative thinking. If there is no inno-vative thinking, there is no graphic design. We should get rid of outdated and simplis-tic habits and constraints of personal intuition and study it by diversification of the imaginary space to implement innovation in graphic design. We can learn and study it from a large number of outstanding creative thinking methods, for example, Dutch painter M. C. features mathematical objects and operations including mathematically inspired woodcuts, explorations of infinity, symmetry, reflection, and tessellations.

For quite a long time, the consumers have been familiar with the application of ink painting in graphic design. In fact, this is a creative behavior for graphic designers, which not only flows from native consciousness but is linked with modern design. Many graphic designers implement ink painting in artistic design to create a new point of creativity. As a consequence, this is a study with innovative think-ing as the starting point. In nutshell, because artistic designing could be viewed as combination of culture and art with science and technology, future possibilities in graphic design must unfold within the boundaries of inherited design. It is necessary to emphasize traditional culture and to positively cultivate traditional design with modern fashion style, which is one of the excellent results of the correlation between creative thinking and graphic design.

5.3.4 Analysis of Creative Thinking in Graphic Design

Innovative thinking mainly refers to the ability to break with the conventional means to achieve innovative new forms of thinking. The innovative thinking of graphic design works is a comprehensive application of different forms of thinking, which is the essence of scientific thinking. It is a comprehensive embodiment between four clusters of thinking – intuitive thinking and analytical thinking; divergent think-ing and convergent thinking; abstract thinking and image thinking; and associative thinking and inspiration thinking.

- *Intuitive thinking and analytical thinking*: The latter, also called logical thinking, should follow very strict logical rules and gradually deduce and implement to get final logical conclusions. The former primarily relies on people's intuition to identify, analyze, and solve problems. It has a premo-nition about the significance and results of a problem or situation to go directly toward the target. Designers naturally combine intuitive thinking with analytical thinking in graphic design. These two factors are in contact with each other and complement each other.
- *Divergent thinking and convergent thinking*: The former expands ideas by thinking from many aspects. It is a new way of analyzing a problem from various aspects, with drawing inferences being a key feature. This is why so many modern people select divergent thinking to solve problems and why it plays an important role in innovative thinking. The latter, mainly based on traditional and single forms to find accurate answers, focuses on in-depth exploration and research in some innovation process. Of course, these two factors are closely related in graphic design.

- *Abstract thinking and image thinking*: Abstract thinking or logical thinking is implemented by conception, judgment, and logical form. What is peculiar about abstract thinking is that it uses abstract conceptualization of visualized content to form concepts, theorems, and theories that sublimate from sensibility to rationality in understanding people and then change from abstraction to rationality. Image thinking shows the reality of visual images with special images and representations of objects as its main thinking form. If we use image thinking in graphic design, it usually makes the work have rich and vivid effects. According to research, there is no innovative thinking if abstract thinking ability and image thinking ability are both weak. Hence, designers should make full use of abstract thinking and image thinking to produce innovative thinking in order to make people satisfied with the works.
- *Associative thinking and inspiration thinking*: AT refers to perceptual imagination, which is an important basic thinking innovative artistic conception. Designers should regularly adopt associative thinking and other thinking forms, for instance, by combining objects and concepts with different meaning as well as other factors that lack an obvious interrelationship. For instance, when you see a cigarette, you naturally want an ashtray, and the same is so for a saddle and a horse. Indeed, this explains the importance of associative thinking in creative thinking activities. Inspiration thinking is produced on the basis of important objects, and it often needs a lot of induced factors. During graphic design, the designer is likely to need inspiration by some object or phenomenon to do good work [3].

5.4   CONNECTING SCIENCE WITH INNOVATIVE GRAPHIC DESIGN

The recent evolution in technological advancements has put scientists and graphic designers on the same page, where the language of graphic design augments scientific design thinking. This has highlighted the need of visual media in communicating the scientifically researched product and findings to the audience. Therefore, Collaborations between researchers, graphic designers, and other visual communications professionals offer great potential to discuss the results of a pilot project and to get their product to the right people at the right time. The need of graphics is not only demanded by journals in giving more impact to research articles and grant proposals but also provides help to viewers to decide in a matter of seconds whether to engage with the material or not.

Researchers and scientists have started accepting the challenge of learning visual media and pushing their work via social media and other online platforms. In this context, graphics people have an added responsibility to collaborate with science to get more tailored product in changing an array of conventional methods and producing a new science media for the audience.

Therefore, design thinking and design science are two ways to innovative graphic designing which aim not only to address individual needs with what is technologically feasible but also to devise a viable product design strategy to derive value from

TABLE 5.1

Design Thinking vs. Design Science

Design Thinking	Design Science
• Explores the cognitive thinking process that is prominent in design action	• Adapts the design process based on the requirement of scientific theory building
• Pinpoints the usefulness of design to deal with socio-cultural factors reframing the problem in a human-centric way	• Builds up systematic order of techniques to be used in design as methodology equipped with formative knowledge
• Deploys human practitioners to develop creative and constructive design practices for solving a problem	• Deploys human and artificial methods to develop an artifact that is efficient and effective to solve a problem

market opportunities. Different theories relating to portrayal of design between relativist art and positivist science suggest that there may be more than one aspect of design as a medium for management research to bring beneficial relevance to domain practitioners. At one extreme, positivist art indicates design science, whereas at the other, relativist art reflects design thinking. The characterization of both of these designing approaches is described in Table 5.1.

In principle, the practice of design thinking and design science respectively feed on the role of research in the multidisciplinary design process. Undoubtedly, it cannot be ignored that both design thinking and design science are derived from design and act as a tool to support the relevance of design in management studies.

5.5 AN INTEGRATED APPROACH TO DESIGN INNOVATION

The essence of innovative graphic design practices is underpinned by the conjunction of design thinking with design science to innovate a solution. These two terminologies have played well in educating designers and practitioners to identify their overlaps in the realm of graphic design and in developing a strong mutual understanding for collaborative solutions to "wicked" problems or to smaller-scale practical ones [4].

The human-centric approach of design thinking makes it important to consider the experience and behavioral values of people while designing a product. Contrarily, design science research looks for implementation of foundation theories, artifacts, and scientific methodology to deal with a practical problem. Having said this, the evaluation methods used in design science research include experiments and field study, in contrast to design thinking, which follows the deep study of human values and feedback for justification and decision making [5].

On the flip side, the two approaches of design thinking and design research hold a common methodology and serve a general purpose of fueling design innovation with research activities. The design thinking framework for innovation can be integrated with design science research activities to achieve the goal of innovative designing

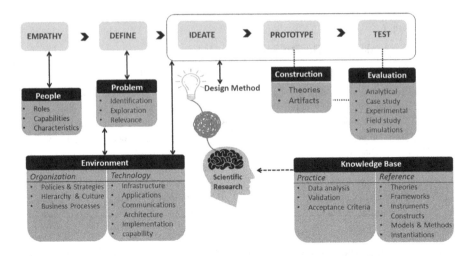

FIGURE 5.4: Design innovation framework, incorporating DSR and design thinking.

with underlying creativity. Figure 5.4 shows a comprehensive view of the integrated design innovation model for graphics and other design disciplines. The figure underlines the rationale behind adapting the three-cycle view of design science research in using the design thinking process in order to develop an innovative framework.

Taking an impression from the three-cycle view of design science research (Section 2.10) and analyzing its relevance with an integrated design thinking approach, it can be said that later stages of the design thinking process utilize the two components of the design cycle in design science research, namely, build and evaluate. Leveraging the methods of construction and evaluation by means of design science research, design thinking professionals can improve design prototyping and testing practices to derive an innovative design solution. As discussed in Chapter 2, design knowledge is created using scientific research and theory, and is then used to test the efficacy of a solution using certain evaluation methods.

Now in the view of an integrated model of design innovation framework, the knowledge base forms a foundation toolkit for design thinking professionals to draw contextual understanding of the problem and wireframe the design outcomes. Capturing applicable knowledge also outlays certain validation criteria that designers need to keep in mind for realization of a real creative design. Thus, imparting the knowledge base and practices of both the worlds in a single model, graphic designing benefits from experience, methods, innovation, and creativity using empirical knowledge and iterative prototyping to reach the final product [6].

5.6 INTERDISCIPLINARY COLLABORATION IN GRAPHIC DESIGN

Gone are the days when people had only an idea but were not equipped with the digital tools and services to convert it into a real product. Over the decades, the advent of science and technology has offered a broad array of software and techniques to understand the "interaction", "service", and "experience" design trends of the market. Nevertheless, the driving force of collaboration among interdisciplinary design disciplines (like graphic design, fashion design, interior design, and industrial design) has been foundational to every successful project. Having said that, it has been observed in recent years that designers are looking to broaden their approach not only by focusing on improving the look and functionality of products but also by performing human-centered research across all the disciplines engaged in creating and evaluating design outcomes that contribute to their product development process. This section of the chapter provides an overview of contextual concepts and opportunities related to this diverse area of activity.

5.6.1 Cross-Disciplinary Approaches – Concept and Opportunities

To meet the unmet needs of today's fast-paced world, designers have to be truly innovative and dedicated toward enhancing the level of quality in their products. This calls for a collaborative cooperation knowledge building and design modeling that connects the theories of experts from different disciplines. In order to explore such opportunities, four distinct cross-domain approaches have been identified in terms of collaboration, as explained below:

a) *One disciplinary*: It is quite clear from Figure 5.5 that this term refers to a single-discipline academic realm that interacts with no other design disciplines. Here, we will refer this discipline as graphic design (GD).

b) *Multidisciplinary*: This approach exists when two or more disciplines work under one umbrella to achieve the vision of an organization. In Figure 5.5, multiple disciplines are represented as graphic design, industrial design, and interior design, along with animation, apparel design, and architecture.

c) *Interdisciplinary*: This approach (Figure 5.6) allows understanding the designers' preconceptions of "what is to be designed" and its realization after analyzing the adopted framework. Essentially, this approach aids in

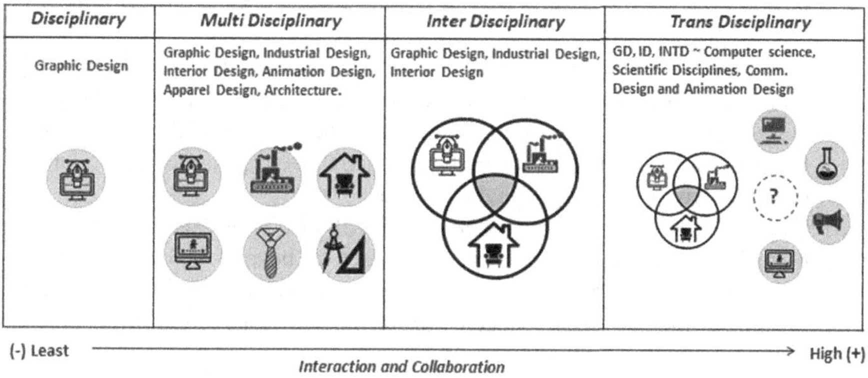

Disciplinary	Multi Disciplinary	Inter Disciplinary	Trans Disciplinary
Graphic Design	Graphic Design, Industrial Design, Interior Design, Animation Design, Apparel Design, Architecture.	Graphic Design, Industrial Design, Interior Design	GD, ID, INTD ~ Computer science, Scientific Disciplines, Comm. Design and Animation Design

(-) Least ——————————————————————————→ High (+)

Interaction and Collaboration

FIGURE 5.5: Cross-disciplinary approaches.

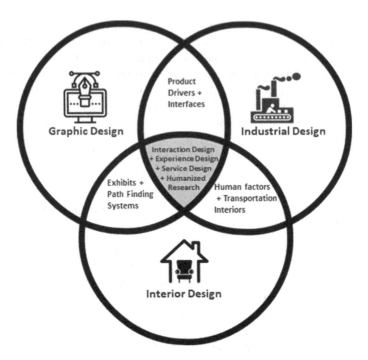

FIGURE 5.6: Interdisciplinary approach.

finding a logical solution to any problem irrespective of the uppermost view and addressing the common areas of concern among diverse disciplines.

In addition to this, the interdisciplinary approach also seeks to find the areas of concern that imbricate learning opportunities in graphic design, interior design, and industrial design. For example, product control interfaces relate to graphic design and industrial design. Demonstrations and path-finding systems concern graphic design and interior design. And physical human factors along with the structure of compact spaces for transport concern industrial design and interior design.

Now there are some interesting areas that are visible at the focal point of all three disciplines. These areas are recognized as interaction design, service design, and experience design. Addressing the basis of these core design fields is a human-centered research approach that enables effective collaboration to determine unique final outcomes.

d) *Trans-disciplinary*: This approach includes all disciplines and tends to optimize some space in identifying new perspectives beyond the limits of existing disciplines. In this system, a large number of disciplines is assembled to address large (or even "wicked") problems than none of the single disciplines could effectively address alone.

For example, dealing with the problem of "big data", or "data analytics", in trans-discipline might need the involvement of the computer science department along with various scientific disciplines that typically generate massive amounts of data through their research practices. In that case, if data is to be transformed into useful information that can be acted upon, disciplines such as communication and design should be included to provide additional expertise. As well, it may be possible that some data can only be properly understood via the use of time-based media. In that case, expertise from the animation discipline might also be required [7].

5.7 TECHNOLOGY FOR INNOVATIVE GRAPHIC DESIGN

The rapid development and adoption of specialized technologies have set the trend for graphic designers to replace traditional techniques with new digitalized tools. Gone are the days when designers in a beginner stage spent a lot of time to gear up on the skills of designing, but now the use of several software tools and advanced technologies has enhanced their creative capabilities in truly realistically rendering their ideas into masterpieces. This digital transformation has also reduced the timeframe of learning with the visualization of products. In this context, science and technology are consistently paving a new path for the smartest minds of the world to introduce a new wave of technology-enabled graphic design that also demonstrates striking features of AI in design. With that in mind, many investors and inventors have now rolled up their sleeves to launch and use the next big crossovers in the realm of graphic designing. Some of the most interesting inventions are described as follows.

5.7.1 ARTIFICIAL INTELLIGENCE

The continuous development in technology and software has made it easier for designers to deliver their creations quickly to the market and reach out to a wider network of target consumers. Additionally, Artificial Intelligence is increasingly bombarding its spotlight across many industries such as computer science, medical, life science, fashion, sports, entertainment, art, and graphic design. Whereas technology has been broadening the scope of designers' creativity in revolutionizing design projects through automation, the use of specific design software for product development is again a promising strategy to discover new opportunities, making the process faster using AI techniques to improve customer experience each day.

Recent trends suggest the advantage of AI as an essential tool for graphic designing. As technology expert and IBM CEO Ginni Rometty talked about the accomplishments of Watson in a Bloomberg Businessweek post, she stated that "AI" would be more helpful to artists if it were understood by designers as standing for "augmented intelligence" and not "artificial intelligence". This statement advocates leveraging AI as a design tool to augment the optimization potential and to speed up the project completions. The power of AI can be used by designers to choose new features that can be projected in existing graphics and well as to work on multiple designs based on constructive feedback that will enhance its market value and sales proposition. However, AI in design technology can never negate the role of human designers in achieving new dimensions of successful creative graphics [8].

How Can Designers Work with AI?

In the era of creativity, collaboration between AI, design, and marketing has proved to be a game-changer. A number of exciting opportunities are still untapped as the designers' world is lagging behind the pace of augmented intelligence. Leveraging such opportunities won't happen until AI and designers' co-creations come into action. Nevertheless, creativity is placed in the crosshairs of science, engineering, art, and design.

AI works wonders for designers to attract more buyers and meet their expectations by tapping their touchpoints [9]. A summary of AI trends in visual design is depicted in Figure 5.6. Airbnb is using machine learning and AI with computer vision technology for design prototyping with production-ready code which aligns with the existing system framework for establishing UI components. In addition to this, Ferrero, an Italian company, has designed seven million prototypes with different color patterns using AI design tool for outreach marketing of its brand Nutella.

5.7.2 IMMERSIVE REALITY

Immersive reality is one of the most popular technologies in design, existing in the forms of Augmented Reality (AR), Virtual Reality (VR), Mixed Reality (MxR), and Augmented Virtuality (AV). These immersive technologies enable interaction with virtual environments by inculcating various interfacing methods such as sensor-based, device-based, tangible, collaborative, multimodal, and hybrid. The acceptance of immersive reality in Human-Computer Interaction (HCI) has offered a benchmark in understanding real-time graphic innovations and providing an exemplary experience of robust, real-time tracking using intuitive interaction interfaces. The evolving trend of immersive reality software in multimedia applications has resulted in an increased utilization of HCI methods and high-performance wearable devices equipped with motion-tracking sensors and image-rendering capabilities. These technologies of immersive reality are nowadays used in entertainment, tourism, education, military, robotics, medicine, and e-commerce industries (Figure 5.7).

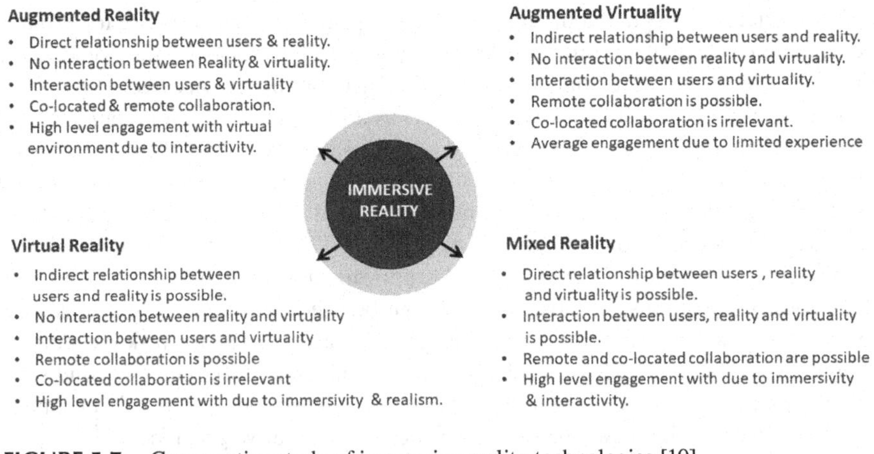

Augmented Reality
- Direct relationship between users & reality.
- No interaction between Reality & virtuality.
- Interaction between users & virtuality
- Co-located & remote collaboration.
- High level engagement with virtual environment due to interactivity.

Augmented Virtuality
- Indirect relationship between users and reality.
- No interaction between reality and virtuality.
- Interaction between users and virtuality.
- Remote collaboration is possible.
- Co-located collaboration is irrelevant.
- Average engagement due to limited experience

IMMERSIVE REALITY

Virtual Reality
- Indirect relationship between users and reality is possible.
- No interaction between reality and virtuality
- Interaction between users and virtuality
- Remote collaboration is possible
- Co-located collaboration is irrelevant
- High level engagement with due to immersivity & realism.

Mixed Reality
- Direct relationship between users , reality and virtuality is possible.
- Interaction between users, reality and virtuality is possible.
- Remote and co-located collaboration are possible
- High level engagement with due to immersivity & interactivity.

FIGURE 5.7:   Comparative study of immersive reality technologies [10].

5.7.3 Sensory Haptic

Unlike the four other senses (sight, hearing, taste, and smell), Haptics – the sense of touch – enables humans to perform a wide variety of exploration and manipulation tasks in the real world. To enhance the performance of humans and to provide stringent control over machines and devices, this technology is used to create virtual objects coded with computer algorithms. Similarly, there are lots of technologies based on Hold, Push, Swipe, and Tap movements that resemble human hand movements in the virtual environment by virtue of digital innovation. But the role of physical entities or tangible products cannot be nullified, as humans and technology have equal impacts on user experience in the event of product purchase. Haptic sense is the perception of paying with "real money" but through transaction in digital space – a feeling of something tangible in your hand that you are giving to someone else, while you are just tapping a flat surface of a screen to navigate payment options.

Today, graphic design activities are carried out in various fields using sensory code technology that creates a haptic sensation of touch by push, vibration, or motion in response to user input even without the intervention of a graphic designer. In general, the design and control of a haptic device are driven by three major haptic systems: *graspable, wearable*, and *touchable*. Figure 5.8 gives examples of each of these types of systems.

1. *Graspable systems*: In these systems, the construction of a haptic device is based on the phenomenon of force feedback, which allows a user to push on the device and experience pushback using a hand-held tool. Another mode of construction in this category involves the use of flywheels or hand-held tactile devices to provide inertial force.
2. *Wearable systems*: The wearable feature of this system is accompanied by smartphone applications to track movements of fingers, thumb, or palm. Here, the haptic device is mounted on the hands and sensation to skin is captured by free in-air movement of the user's hand. The haptic response may provide cues such as vibration, lateral skin stretch, or normal skin deformation. Some body-grounded devices such as exoskeleton can also be used to obtain kinesthetic cues by creating a reaction force on a less sensitive part of the body.

Graspable   Wearable   Touchable

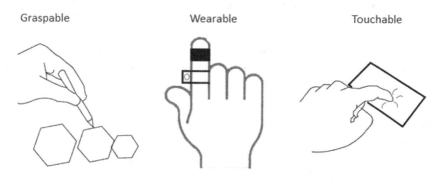

FIGURE 5.8:   Types of haptic devices [11].

3. *Touchable systems*: These systems are encounter-type displays or smart kiosks that allow the user to actively explore the entire surface. They can be purely skin-responsive devices that change their tactile properties based on positioning of touch input. Hybrid touchable devices are able to change their shape, mechanical properties, and surface properties.

The applications of haptic-enabled graphics are mostly found in mobile communication and AR gaming. In addition, sensory haptics is also making its way into healthcare and medicine where the sense of touch is critical to the situation (e.g., remote surgery and prosthetic limbs). The commercial use of haptic devices is limited by the availability and responsiveness of embedded hardware, and thus leads researchers to use ready-made components to build their own haptic devices. For any design process for human-machine interaction, the operational performance, expressiveness, and effectiveness of haptic hardware to sense various touch inputs are the foundation of success.

SUMMARY

Overall, Design Thinking and Design Science are the two key performance indicators of innovative thinking that not only play an important role in designing innovative graphics equipped with different technologies but also are instrumental to the commercial economic growth of a state or a nation. Therefore, one should prioritize developing skills in design research and design science and building innovative thinking ability by utilizing the formal language of creative graphic design methods. Both aspects owe their origins to "design" and aspire to solving problems in the sphere of practice.

At its outset, this chapter has shown that design thinking and design science are complementary components of an overall design paradigm where design science is the primary methodology of choice for research of better-defined problem areas and design thinking is especially suited to the more wicked type of complex problems, where abductive reasoning, intuition, and creative thinking and methodologies are necessary, which are based on internal, tacit knowledge of the (co-)designers.

Thorough understanding the interdisciplinary approach of innovation and creativity in graphics, designers can set a new dimension in reaching out and meeting the needs of prospective users. In the era of science and technology, designers are leveraging various technologies equipped with artificial intelligence, augmented reality, and immersive media in achieving new dimensions of successful creative graphics. As outlined in this chapter, graphic designers have been encouraged and motivated to adopt a new wave of technology which is consistently paving a new direction to bring more value and satisfaction to their clients.

REFERENCES

1. 5 Steps of the Design Thinking Process: A Step-by-Step Guide, Voltage Control. Available at: https://voltagecontrol.com/blog/5-steps-of-the-design-thinking-process-a-step-by-step-guide/

2. Ian Noble, Russell Bestley. *Visual Research - An Introduction to Research Methodologies in Graphic Design*, AVA Publishing SA, 2005.
3. Yingying Ma. "Innovative Thinking Plays Important Role In Graphic Design", *Advances in Intelligent Systems Research, International Conference on Education, Management and Computing Technology(ICEMCT 2014)*, Atlantis Press, 2014: 111–114.
4. "Wicked Problems: Problems Worth Solving", *Austin Center for Design*. Avaialble at: https://www.wickedproblems.com/1_wicked_problems.php
5. Joseph Giacomin. "What Is Human Centred Design?", *The Design Journal – An International Journal for All Aspects of Design*, Taylor & Francis 17(4), 2014: 606–623.
6. Frank Devitt, Peter Robbins. "Design, Thinking and Science", *Communications in Computer and Information Science, European Design Science Symposium* 388, 2013: 38–48.
7. Paul Nini. "Interdisciplinary Collaboration in Design: Context and Opportunities". Available at: https://medium.com/@pjn123/interdisciplinary-collaboration-in-design-context-and-opportunities-92e63828a4ed
8. Alan Hevner, R. Alan, March Salvatore, T Salvatore, Park Jinsoo and Ram Sudha. "Design Science in Information Systems Research", *Management Information Systems Quarterly* 28(03), 2004: 75.
9. Available at: https://www.toptal.com/designers/product-design/infographic-ai-in-design
10. Available at: https://www.frontiersin.org/articles/10.3389/frobt.2019.00091/
11. Heather Culbertson, Samuel B. Schorr, Allison M. Okamura. "Haptics: The Present and Future of Artificial Touch Sensations", *Annual Review of Control, Robotics, and Autonomous Systems, AS01CH12_Okamura ARI* January 2018. doi:10.1146/annurev-control-060117-105043

6 Sixth Sense
Panoply of Design in Digital

6.1 INTRODUCTION

Technology is the means by which humans tend to catalyze natural processes to meet their changing demands. An increase in the requirements for a new process leads to an increase in the development of technology that can address those requirements. Nowadays, technology has almost erased the word "impossible" in its multifaceted realm. Every day, new inventions, discoveries, and improvements are making this evidence stronger while amazing people by solving complex tasks ever more quickly. As the current market continues to be inundated with the bulk of technology advancements, it is very difficult to recognize the next big thing for digital users.

Sixth Sense Technology is a novel language of computing that combines the physical world with big data or digital information. This technology has blazed into prominence in the hi-tech sector and exhibits its relation to the capabilities of six senses. By definition, the technology behind "Sixth Sense" is a wearable gestural interface that lets humans use their natural gestures to communicate with the information embedded in hardware like mobile devices. In other words, Sixth sense technology seeks human-computer interaction by utilizing the principles of computational and behavioural science.

The five senses, when faced with a new experience, try to capture the stimulus and its effect. This capture is further analysed and used by the sense organs to interact with the environment. Having said this, there are many tiny handheld computing devices that also help people to interact with the digital world from anywhere in different situations. However, with changes in psychological and behavioural intent, more people are demanding the next possible use of technology to make the ends of both the real world and the digital world meet in a way that benefits them in multiple aspects.

Changing the way we use mobile devices to gain information from the Internet, sixth sense technology facilitates the availability of important information and access to the Internet using a device that is no bigger than existing mobile phones; surprisingly, it is even available in devices the size of a small shirt button. Besides the foundation of scientific principles, sixth sense makes use of augmented reality to integrate physical objects with the digital world. The scientific definition of sixth sense is the power of reception and perception of information seemingly exclusive of the five, which are seeing, hearing, smelling, tasting, and feeling.

6.2   HISTORY OF SIXTH SENSE

Sixth sense was not recognized as a term during initial developments until 2001, when Steve Mann postulated his work to describe such devices and named them "Sixth Sense". Steve chose this name for the reason that wearable computers could receive and provide information in addition to the five bionic senses. Thereafter, it was implemented by Pranav Mistry, an Indian researcher at MIT. Steve Mann is known to be the father of sixth sense, as he made a startling invention of a wearable computer in 1990. The model of sixth sense has its roots in the first-of-its-kind head-worn gestural interface developed by Steve Mann at MIT Media Lab in 1994. This system was quickly revised, including a neck-worn gestural interface which was also developed in 1998 by Steve Mann.

The latest development in the sixth sense applications we see today is the result of an innovative digital sensor developed by both Pranav Mistry and Steve Mann, who collaboratively created hardware and software for sixth sense to work in 2009. Holding different contexts in perception and understanding, the term "sixth sense" has been used by different inventors in diverse fields to describe versatile digital capabilities enabling users to access technology for reading and sharing information with those around them.

6.3   DESIGN PRINCIPLES OF SIXTH SENSE

The steady evolution in the field of technology has helped people in realizing their dreams of a better future and becoming stronger every day. Numerous technological developments have made human life easier, but now is the time to pause and rethink the future. Surrounded by many digital devices capable of sending and receiving signals, people have seemingly gone away from the real world. Although these devices, mostly portable, engage them in consuming apps and games that entertain their mind and also help them to do work, many such digital devices that can adapt to real-time situations are still far from the reach of people. Taking them away from their natural environment, most devices that easily scale to size of a palm are adversely affecting the minds and bodies of their users.

To make a device which has the power of perception, reception, and action is the new normal in wearable technology applications. Using this as a benchmark, designers and engineers have developed numerous interfaces that are able to receive information from the surroundings and interact with people. Sixth sense serves this purpose by sharing secondary information supplied through a smart wearable device when it connects with a person.

Sixth sense works on the foundational principles of gesture recognition and image processing. It makes use of visual and sensory components that bridge the gap between the logical and physical realms of digital information and allows people use their hand gestures to interact with this information in a projected interface. Moreover, sixth sense devices support multi-user and multi-touch interactions. The design model of a sixth sense device is described in the following section.

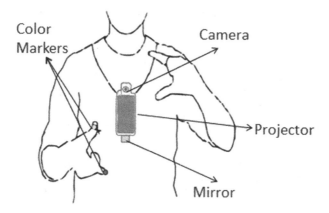

FIGURE 6.1:   Components of sixth sense device.

A sixth sense device (Figure 6.1) operates on two instrumental technologies: gesture recognition and image processing. To enable sixth sense technology, the device prototype is made up of easy-to-access components like camera, projector, mirror, color marker, and smartphone. The entire sixth sense device is fabricated in the form of a wearable pendant.

- *Camera*: A Camera tracks the motion of color markers using programmed algorithms.
- *Color Markers*: Segments of Colored tape are wound around the fingers of the user to communicate during tracking process.
- *Projector*: A Projector tracks the location of the color markers and stores the data on the projected surface.
- *Mirror*: A Mirror reflects the image processed by the projector to the front of the user's view.
- *Smartphone*: A mobile device like a smartphone transfers and receives data and voice signals using the Internet.

WORKING OF SIXTH SENSE TECHNOLOGY DEVICE

1. The projector and the camera are connected to the mobile phone. The projector projects the visual images on surfaces like a wall, a table, or even one's palm. These surfaces act like an interface for communication.
2. The camera recognizes the gestures through physical objects and movements made by user with colored markers (red, blue, green, and yellow) using image vision techniques.
3. A software program processes the course of video acquired by the camera and maps the gestures. Different gestures are assigned specific commands based on the movement and the direction.

There are three broad gestures recognized by software in sixth sense:

- *Framing Gesture*: this gesture captures the image of a scene. The user doing a framing gesture can stop by a projected surface and flick through the images taken.
- *Multi-touch Gesture*: This gesture resembles the touchscreen feature where a user can move the map by dragging and pinching movements.
- *Freehand Gesture*: This gesture depicts an event when a user draws an icon or object form in the air.

Mobile devices equipped with software allow searching on the Internet for information that is relevant to the current situation. In this way, a sixth sense technology–enabled device can determine not only what a user is doing but also how he is communicating with the physical world using digital information with sixth sense.

6.4   TECHNOLOGIES RELATED TO SIXTH SENSE

- **Augmented Reality**
  Augmented reality, one of the most promising visualization technologies, is suitable for enabling humans' limited perception capabilities to implement digital objects into the real environment. This blends visual overlays, sensory projections, haptic feedbacks, and sounds into the natural environment as it exists. The use of augmented reality in sixth sense devices allows users to view the elements of the physical world around them through smart sensory devices (like cameras) while projecting onscreen objects and viewing them as if these are right in their real-life environment.

  Augmented reality requires the user to have a device through which he will input information by typing or touching the screen. But when it collaborates with sixth sense, a user can easily feed information as quick as his thoughts and finger swipes. The three essential components of a sixth sense augmented reality system are the tracking device, the head-mounted projector, and the mobile computing hardware. All these components collectively pinpoint the user's position in reference to his neighborhood and track his eye and head movements.
- **Computer Vision**
  As a technological tool, computer vision in sixth sense makes use of a computer that can extract useful information from an image or multidimensional data to generate human-readable output. A computer vision system is built upon the image-processing algorithms and feature detection techniques to solve a given task. Computer vision seeks to study physiological processes resulting in visual perception capabilities in humans. It also describes the techniques for developing software and hardware in computer vision systems.
- **Gesture Recognition**
  This technology is built to interpret human gestures using computational algorithms. The gesture recognition technique interprets the emotion

captured from the face-hand gestures. This helps a user to interact with the computer directly through text and graphics without the need of external peripheral devices for interface, but requires precise tracking of hand movements. Gesture recognition can be well thought of as a one-of-its kind technological innovation that understands human gestures and resembles human-computer interaction through a cursor pointing to the framing symbols. Adaptive models [17], neural network prototyping [22,23] and statistical modeling are commonly used for enabling gesture recognition. In object recognition using RFID, devices are tagged to interact with each other through radio signals.

6.5   COLOR SEGMENTATION IN SIXTH SENSE DESIGN

Compared to other electromechanical devices, sixth sense devices are more intuitive and powerful. Color segmentation plays an important role in the performance of a sixth sense device as it uses hand gesture recognition by means of color markers to perform operations. A color space model detects and segments various colors from captured images in real time. The processed image is comprised of segmented color pixel values and neighborhood values between two colors. The following sections explore the color segmentation technique [12] by Shoaibuddin and Ravindra, which is based on the foundational working of the RGB model and MATLAB®.

Color recognition is an important step for recognizing gestures. The foremost step is to manipulate the video frames into images that consist of three color components: Red, Green, and Blue. The video from the camera is converted into frames with defined pixel values. These frames are further converted from the RGB format to a different color format, as a part of RGB color normalization. The gestures are input by using color marker caps; therefore the recognition and segmentation of color become significant (Figure 6.2).

After the color segmentation, the location of the centroid should be calculated. The Euclidian distance between the centroids depicts the corresponding gesture. The workflow of color image processing in sixth sense devices can be learnt in three stages: color recognition, actuation, and color segmentation. Here, MATLAB is used for instantiating image processing, color recognition, and color segmentation.

a) *Color Recognition*: The Color recognition system seeks to find a person wearing red, green, blue, and/or yellow color markers. This methodology captures multicolor images, and the area of interest within an image is selected using MATLAB command. The selected portion's pixel values are extracted and color information is stored using a calibration method. The process converts the video into a number of frames and each frame value is computed with the stored pixel values using Euclidean distance formula. The Euclidian distance is calculated using the following formula: $E = \sqrt{(XX2 - XX1) - (YY2 - YY1))^2}$

b) *Actuation*: The colors displayed on the camera are acquired and compared with the saved pixel values. Different colors are recognized based on the description of pixel values and then presented on the screen.

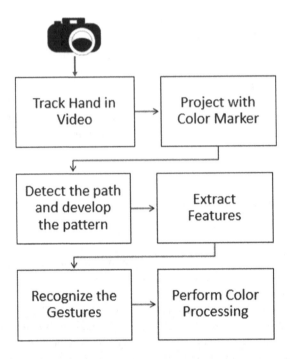

FIGURE 6.2:   Stages of color-processing framework.

c) *Color Segmentation*: The model utilizes RGB space normalization for color segmentation. Normalization of color is useful for removing the effect of light and shadows on the object. This methodology makes use of Euclidean distance formula to match or compare the distance of the original pixel values with normalized pixel values. After segmentation, the centroid of each color segment is calculated, and further the distance between number of centroids of specific colors is calculated.

The above three operations use image-processing techniques and can be simulated using a MATLAB or LabVIEW setup. An optimized color processing for a sixth sense device can filter noise and reduce design complexity with better hardware support.

6.6   APPLICATIONS OF SIXTH SENSE TECHNOLOGY

Sixth sense is a wearable technology that is portable and supports a multi-touch user experience. This technology can be used by anyone without access to a keyboard, camera, or mouse for capturing and exchanging any sort of information in the digital world. The following are some of the many interesting applications of sixth sense that are making it easy for people to access information insights from anywhere at any time.

1. *Picture Zoom Feature*: This is the most commonly used feature used in online reading or a case study. The scaled-down text of the e-books can be enlarged using zoom feature with finger movements. By the use of sixth sense technology, a reader can take advantage of enhanced readability and deep visualization of the content in understanding the topic of interest, without using a computer (Figure 6.3).
2. *Palmtop Calculation*: Using the sixth sense technology, a keypad can stick out on our palm such that numbers are visible on the fingers. This keypad can be used as a calculator itself with the help of numeric buttons. This makes the portability feature useful to students without requiring them to carry calculators or mobile devices all the time (Figure 6.4).
3. *Immersive Multimedia*: The most striking application of sixth sense is that it can convert a piece of paper into motion media. For example, while

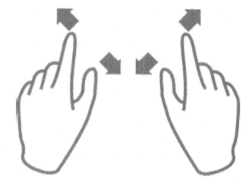

FIGURE 6.3:   Screen zoom using fingers.

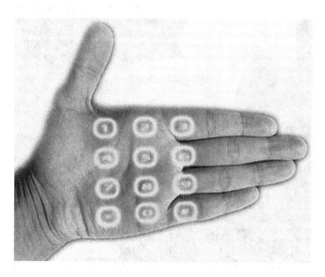

FIGURE 6.4:   Illustration of keypad on palm.

reading the column, the real-time information can be projected onto the content without the Internet. A sixth sense device can recognize articles or relevant stories from the Internet and display them on pages before the user. For book lovers, sixth sense technology comes with a bonus feature of converting readable content into audible text.

4. *On-Demand Phone Conferencing*: Sixth sense technologies make it possible to project a keypad on the palm and make a quick call by dialing numbers without needing to carry a mobile phone.

5. *Capturing Live Images*: Using framing gestures, an individual can capture scenes or particular areas while traveling and save them by virtue of sixth sense technology. The captured images can be projected onto the desired surface and can be sent, organized, or sorted using hand gestures (Figure 6.5).

6. *Virtual Environment*: This application enables the user to establish an artificial class or lab anywhere at any time without access to the real components of the computer. The tabletop can be converted into a monitor screen and the user can surf data files or software without having physical peripheral requirements. The idea of a virtual classroom using sixth sense outlines the role of blackboard teaching methods online, for example, e-learning. By integrating smart classes with sixth sense technology, tutors will be able to offer interactive lectures. This facility can also be utilized to create and publish study journals in digital format (Figure 6.6).

7. *Smart Object Tracking*: This is the most advantageous application of sixth sense for tracing objects in a situation when you have security cameras with sensors installed at your home. You can send a query like "Find me clips of video recording when I was away". This would return a number of clips from records of corresponding events, making use of sixth sense.

FIGURE 6.5: Taking picture using frame gesture.

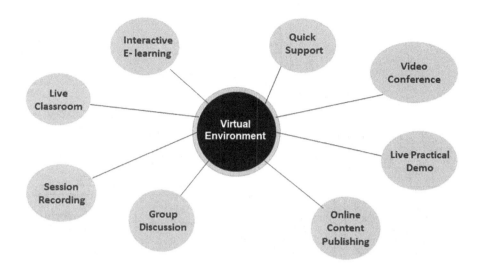

FIGURE 6.6:   Applications of virtual environment.

6.7   GESTURE-BASED HUMAN-COMPUTER INTERACTION

The word "gesture" refers to a nonverbal means of communication in which an individual uses bodily actions like the movement of hands, head, or face to communicate different sorts of messages with or without speaking. Gesture is interchangeably used with posture, but gesture is a broader term that provides more information, including body language, movements, and facial expressions. A gesture [1] is a static or dynamic orientation of the human body to communicate ideas and emotions. Static gestures include pointing a finger to refer to something; a victory sign, a v-shaped form using the middle finger and the forefinger; or two palms joined together to show respect and gratitude. On the other hand, dynamic gestures essentially include motions or movements to express certain ideas or information; for example, the head and hand gestures of an umpire to communicate decisions during a game. Combining static and dynamic gestures, a complex gesture is also commonly viewed in dance forms, which include a mixture of static and dynamic symbols, facial expressions, and body language.

In addition to static and dynamic forms, gestures are also classified based on the degree of communication with a second person [3]. If the communication is not intended for others and is subjective to oneself, then the gesture is an inherent gesture. This gesture is most commonly formed by individuals while looking for better understanding of text while reading an article or chapter, or thinking deep about a solution to some hard problem. One inherent gesture, finger biting, expresses the state of mind of a person. In contrast to inherent gestures, outward gestures are also useful in communicating negative and positive emotions. Outward gestures for signaling or instructions include palm and finger movements by traffic police or by a teacher asking his student to stand up or do something (Figure 6.7).

Cadoz [7] suggests that gestures can be classified based on their functions as semiotic, epistemic, and ergotic gestures. Karam and Schraefel [9] further investigated

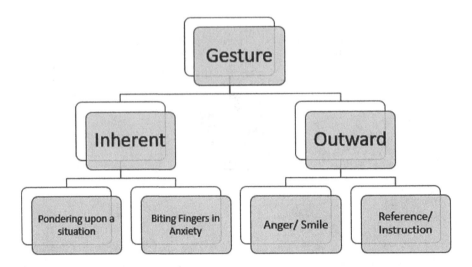

FIGURE 6.7: Classification of gestures.

a classification taxonomy that relates to the multidisciplinary research field of HCI technology. They tailored the classes of gestures into deictic, depictive, manipulation, semaphores, and sign languages.

a) *Deictic gestures*: These gestures point to objects and directions without necessarily stretching the index finger, and can be performed by using the thumb, the other fingers, or the palm.

b) *Depictive gestures*: These gestures are iconic gestures which are used to indicate physical objects referred to by use of speech.

c) *Manipulation gestures*: These gestures signify guided movement in a feedback loop and establish a relationship between the user and the object to be manipulated. Here, the actor or the guide waits for action to be followed by another entity on the current instruction, before proceeding to the next part.

d) *Semaphores*: Semaphores are comprised of shapes and motions of the hand that convey a learned meaning. Semaphore gestures seek for the background experience of the user to communicate with the environment.

e) *Sign languages*: Sign languages are characterized by their high level of learning requirements in order to push a gesture communication. These languages are often complicated and hard, requiring a deep expertise to design for the HCI system as well.

As is true for any gesture language, it should be designed and developed through deep research of multiple user experience, their understanding and preferences of using gestures in the real environment.

6.8 GESTURE INTERACTION TECHNIQUES

1. Gesture Acquisition

   With the rapid development of social platforms and digital economy, the demand for the human-machine interactive technologies is being greatly enhanced. Among these technologies, hand gesture acquisition is a prominent choice to collect and process human gestural information. Gesture tracking and acquisition use specially designed cameras or sensors having a range of visible to larger wavelength and wider spreading. The most widely used instrument for gesture tracking is the Kinect system [2] developed by Microsoft, which operates on a set of motion-sensing input devices (infrared sensors, RGB camera, and detectors) to map dynamic gestural information. Kinect can be used as hands-free user interface to interact with the computer and is positioned on top of the display device like a webcam (Figure 6.8).

   The motion-tracking technology of Kinect is powered by its depth-sensing capabilities using multiple light patterns. The infrared source in Kinect projects the light, which is captured by an infrared sensor. Infrared light reflects off the neighborhood objects, enabling sensors to track the deformation of modulation pattern by pixels. The time of flight for closer objects happens to be shorter, and this allows better accuracy in calculation for more frames per second. Kinect demonstrates state-of-the-art machine learning features, which make it one of the most preferable potential systems for enterprise software and artificial intelligence applications (Figure 6.9).

2. Gesture Recognition

   Present hand gesture recognition technologies are dependent on visual data or sensor hardware; the former uses the hand as an input peripheral and offers an uninterrupted, natural human-machine interactive experience, and the latter uses sensor hardware, which is limited by its cost of implementation and other unfavorable constraints [5]. Not surprisingly, vision-based hand gesture recognition is a potential technology that allows a user to leverage realistic interactive interfaces with freedom of motion. Hand gestures form a basic biometrics feature and are of paramount significance as a key recognition technology for intelligent human-computer interaction [6,10].

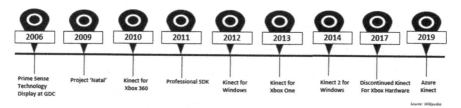

FIGURE 6.8: Timeline of Kinect development.

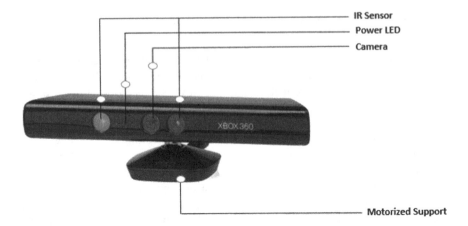

FIGURE 6.9: Working model of Kinect sensor [3].

Factors influencing gesture recognition include the following:
a) *Modality*: Gesture recognition in sixth sense devices is fundamentally performed using two-dimensional and three-dimensional modalities [3]. A two-dimensional gesture recognition approach relies on computer vision methods to extract the recorded movements of the human body, whereas a three-dimensional gesture recognition technique uses sensors and markers to obtain the movement track with the help of depth images. An independent interface that enables the interaction of a machine with a human is called a modality [8].
b) *Relative Change in Position*: Gesture recognition with regard to a change in the positioning of hands can be classified as static gesture recognition or dynamic gesture recognition. A static recognition method generates the relative position of the hands and arms at specific points of time. Static gesture recognition processes only one image and does not require path information to segregate its features. Contrarily, a dynamic recognition method [6] requires a complete sequence of gesture frames to be tracked and the relative shift in the trajectory of motion to be extracted over a certain period of time.

6.9 MACHINE LEARNING TECHNIQUES FOR HUMAN GESTURE INTERACTION

To support unconstrained interaction procedures, machine learning is the most sought-after technique to recognize gestures from human body postures and the movement of features that demonstrate emotions. Over the years, there have been classical studies on perception of human gestures and their interpretation. Machine learning as a technology solution to gesture recognition problem areas has become a crucial need of the hour to realize intelligent human-computer interaction. Reading

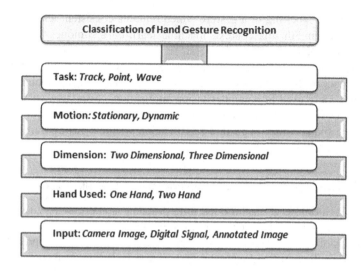

Classification of Hand Gesture Recognition

Task: *Track, Point, Wave*

Motion: *Stationary, Dynamic*

Dimension: *Two Dimensional, Three Dimensional*

Hand Used: *One Hand, Two Hand*

Input: *Camera Image, Digital Signal, Annotated Image*

FIGURE 6.10: Hand gesture recognition techniques [14].

the above, it is important to design a gesture interaction model that can intelligently interpret human body movements and is able to perform high-level tasks involved in spatial interactions (Figure 6.10).

In recent years, continuous research efforts focused on gesture tracking and motion analysis experiments have led to avant-garde machine learning techniques and advanced image-processing sensor cameras to execute gesture recognition. This has further opened opportunities for the development of improved HCI interfaces and gesture interaction applications in many industry domains.

6.9.1 RULE-BASED APPROACH VS. MACHINE LEARNING APPROACH

A rule-based approach to design interaction model relies on a set of threshold values and metrics to track the motion of the object. Using this approach, developers specify a set of rules for gesture description framework. The advantage of applying a rule-based approach lies in the developer's ability to detect the gesture by coding in a high level of abstraction. The Flexible Action and Articulated Skeleton Toolkit or FAAST (Suma et al. 2011) supports gesture creation based on the selection of threshold values for calibrating gestures. This uses OpenNI-compliant depth sensors to interface between human body control and augmented reality applications. More usefully, the FUBI framework designed by Kistler in 2012 uses XML-based definition language to describe complex human gestures. A rule-based approach requires intensive developer efforts in creating a set of rules for gesture recognition in a highly dynamic environment such as recognizing the direction and track of the body movements of a player making different shots.

This limitation is overcome by artificial simulation methods like supervised machine learning where the gesture recognition is referred as a classification problem to address the situation. Gesture identification as a classification problem is characterized by labelling of a gesture which remains consistent with existing data about the problem. In this context, machine learning techniques leverage model training sets to generate a unique classifier label for each gesture. This classifier further examines the likeliness of trained gestures with a natural gesture, resulting in the label of the most resembling gesture. Machine learning techniques using Kinect are widely used in computer vision applications. Similarly, gesture interactive interfaces built upon machine learning algorithms address the problem of the performance and accuracy of gesture classification system by using training datasets of gesture acquisition.

6.9.2   Framework of Machine Learning–Based Gesture Recognition

Gesture description as a machine learning problem is addressed by two stages of training and recognition [11]. The training approach feeds on recognition techniques with samples of various gestures to be identified. The recognition approach follows training and applies a trained set to detect the natural gestures performed by the user.

1. Training Stage
   This stage involves the user acting as a trainer and performing hand or body movements in front of a system (e.g., Kinect). The system arranges the performed movements in sequence and stores them in a target gesture training dataset. The procedure is repeatedly executed to retrieve numerous but meaningful samples of gestures to be recognized.

   After the gesture training dataset is obtained, center transformation and normalization are performed to improve the accuracy of gesture recognition as a result of differences in body poses and positions within a detection field. The center transformation shifts the sequence of gestures to the center of the coordinates in three dimensions. The process of normalization examines the relative distance and multiplies all the positions of body joints by a scalar quantity. Once the machine modeling techniques are trained, the gesture recognition is performed in the next step.

2. Recognition Stage
   At this stage, a user is viewed as an actor performing certain trained gestures before the system. Gesture recognition through machine learning seeks to evaluate the performed gestures and matches with trained ones. Based on the comparative analysis, the recognition approach results in the presentation of recognition outcomes to the user.

6.9.3   Computational Methods for Gesture Recognition

1. *Dynamic Time Warping*: DTW or Dynamic Time Warping [12] is a temporal pattern-matching algorithm which is widely used in human speech and text recognition. This algorithm measures the distance between two objects of a sequence and aligns them until an appropriate match between these two objects is plotted. For gesture recognition, the DTW algorithm takes

data from recently performed tasks. Any movement by a user is stored in a data buffer and compared against a reference trained data buffer using a DTW algorithm. The elements of the current data buffer are represented in Cartesian coordinates to calculate the distance pattern using the Euclidean distance formula. The Cartesian coordinates of the recognition data buffer are adjusted to align with the user's requirements.

The DTW algorithm offers extended versatility and greater possibility of hits for gesture recognition, especially when the gestures are performed at different speeds. The computational performance of the DTW algorithm is influenced by its number and length of reference data patterns. The number of computing iterations required to measure the distance is proportional to the length of the gestural patterns.

2. *Hidden Markov Model*: A typical human gesture recognition system compares a sequence of body positions over time to a number of sample gesture sequences, each of which are trained using a reference dataset. If the sample description seeks for experimental interpretation, the DTW algorithm is combined with a hidden Markov model (HMM) technique. HMM [13,14] is a statistical Markov model in which the system being trained is taken as a Markov process which is a function of X and Y. X refers to hidden or untraceable states, and Y refers to a recognition parameter whose value depends on X. The goal of HMM is to obtain information about hidden states (X) by observing the varied patterns of Y. HMM attempts to convert gestures into a sequence of symbols instead of applying geometric coordinate features. Rich in spatial-temporal object recognition capabilities, a two-stage HMM model proves very effective for applications in gesture-controlled human-robot interaction [15,19].

3. *Naïve Bayesian Classifier*: The foundation of naïve Bayes gesture recognition is based on three main factors: (1) the effectiveness of the naïve Bayesian classifier, (2) the relationship between linguistics and machine vision, and (3) the cumulative-proof inference process of a naïve Bayesian network [16]. A hybrid Bayesian network can be applied to various image-processing applications, which proves its vitality in embedding images and objects of HMM and thus allowing different types of extensions for temporal and behavioral structures [18]. In comparison with HMM, naïve Bayes classification represents dynamic gesture recognition with a small number of data metrics, and that too without disturbing the natural pattern recognition rate of the model.

6.10   UNDERSTANDING SIXTH SENSE ROBOTICS

The path of interaction between man and machine has been developing using various computing technologies. Sixth sense robotics is a predominant field of application of sixth sense in designing the mind of a robot. The study of interaction between human and robot utilizes the foundational technology of gesture recognition and draws contributions from multidisciplinary fields of artificial intelligence, design, natural language understanding, human-computer interaction, robotics, and social sciences. The three laws of robotics stated by Isaac Asimov in his novel *I, Robot* (1941) provide an ethical basis for the safety of humans and proclaim the use of efficient technologies to detect barriers in the protection of humans from potentially dangerous robotics equipment.

Robots are digital agents who can perceive and act in the physical world. In recent years, their use has been found in the most technologically advanced domains of search and rescue, defense, bomb detection, scientific research, hospitality, entertainment, and healthcare. These domains imply a closer contact with humans. Having said this, robots and humans will always share common goals of achieving certain tasks which ultimately require exploration of theoretical models and evaluation.

6.10.1   HUMAN-ROBOT INTERACTION

Human-robot interaction is a scientific and technological research field dedicated to gaining knowledge about the design and evaluation of robotic systems to be used for human applications or to be used with humans for integrated system development. An essential element of the human-robot interaction is the nature of communication [8]. Major modalities used in human-robot control systems need prior training, which can be time consuming. The goal of studying human-robot interaction is to develop a way of easy communication with the robot through human gestures, voice, and facial expressions.

The term "robotiquette" defines social rules for behavior of robots that is collaborative with and comfortable for humans. A common approach to program social etiquette into robots is to first explore the style of social collaboration of human-to-human interaction and then transfer that knowledge in the form of training. The application areas of human-robot interaction include smart robotic technologies that are used by humans to perform industrial, medical, space, social collaboration, and other multipurpose tasks.

* *Space Robotics*:
  Aiding several breakthroughs in aerial and space exploration, space robotics has played an important role in assisting astronauts with specialized surface-penetrating radar and planetary rovers that operate under extreme conditions. Information gathered by the space robot is communicated to the astronaut in close proximity or to the ground-based teams, who then evaluate real-time patterns to improve task performance by changing the behavior of the robots.
* *Industrial Robotics*:
  Industrial robotics deals with the interaction of robots with humans to execute production tasks in collaboration. While the cognitive intelligence of humans allows them to choose different approaches for problem solving, they can choose the best fit among all available options, and then program robots to perform assigned tasks. Taking advantage of selected choice by humans, robots become more consistent in performing repetitive actions and prove their efficiency in manufacturing processes. However, this might

put people in danger who are working with robots in the same workspace for the reason that robots have the capability to move and operate sharp, heavy tools quickly with force.

- *Healthcare Robotics*:
  Medical robotics in healthcare leverages assistive robot systems to provide physical, mental, and social health support, especially to the elderly and disabled. Being an important application area of human-robot interaction, healthcare robotics implies friendly robot interaction and safe physical contact in proximity with disabled persons. Mapping human cognitive intelligence to render natural interactions such as gestures and speech, robots are designed with varying structures like animal-faced robots, wheelchair robots, manipulators, and humanoids. Factors like closeness and long-term human interactions influence the performance of assistive healthcare robots.

- *Atmosphere Sensing Robotics*:
  Robots have been used in assisting search operations and rescue operations. Remote sensing robotics works directly with individuals, most often with trained personnel, for disaster rescue operations. In this field, a small robot is designed to be used in potentially dangerous areas to search for victims hit by mistimed disaster events. These robots are equipped with environment-sensing equipment and video cameras to perform their tasks. The domain of atmosphere sensing for search and operation is innately unstructured, but offers a rich experience of human-robot interaction to locate victims and explore their surrounding conditions.

- *Automotive Robotics*:
  The most recent use case of robotics in automotive domain is the robotic driving automation. The goal of this human-robot vehicle collaboration is to ensure the comfort and protection of drivers using automated systems. The continued progress in improving the working of the system toward autonomous vehicles aims at making a more efficient driving experience that is safe and does not require human intervention to handle braking and acceleration in an unexpected situation while driving.

6.10.2 SIXTH SENSE ROBOTICS IN DRIVER ASSISTANCE SYSTEMS

Enrique Coronado et al. [8] proposed an open source gesture-based robotic operation framework for a car-like vehicle control system. The framework uses human gestures as the interaction interface and identifies a suitable association of gestures with corresponding motion control commands. The principal objectives of offering this framework as a benchmark have been accepted across the globe by vehicle manufacturers to build autonomous driving assistance systems.

- **Objective 1:** *Full control of the robot's movements*
  This is achieved by controlling the moving velocity and angular velocity. Meanwhile, a continuous map is generated to track the gesture-induced data and the state of velocity pair.

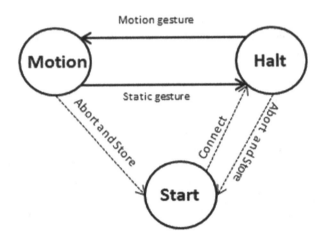

FIGURE 6.11: Functional state transition.

- **Objective 2:** *Full control of the states of robot's functions*
  A simple finite-state machine approach to achieve this objective is represented by three states: start, halt, and motion. For zero speed, no sensory
  data is accumulated (Figure 6.11).

Research in the field of moving object interaction is dominated by the gesture-based
vehicle control systems used by the automotive industry. They comply with detailed
design guidelines as instructed by international standards of governing bodies, such
as the European Commission. Donald Norman et al. [21] suggests six fundamental
principles for a usable robot interface:

a) Simplify the task structure
b) Make control options visible
c) Correctly map the pair of gesture movement and velocity
d) Outlay the power of constraints
e) Allow design components for errors
f) Make the utility clear

The interactions between human and robot are classified into remote and neighborhood interactions [20]. As the name suggests, remote interaction seeks for communication between human and robot when they are remotely located, separated by space
and time. In contrast, neighborhood human-robot interaction allows communication
when these two entities are placed close to each other.

Application of Robot with Sixth Sense

Over time, the interiors of cars have been modified to include advanced features
built on advanced technologies. These technologies, such as sixth sense robotics,
gesture recognition, and speech processing, are installed in the car control systems
to assist drivers in various conditions. To meet safety measures and the drivers' comfort needs, advanced driver assistance systems (ADAS) [23] are increasingly being

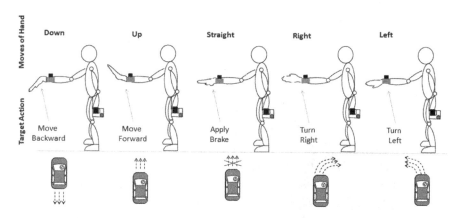

FIGURE 6.12: Functions of driving assistance robot.

launched in the automotive design industry. These ADAS are equipped with technology underpinning the behavior of interfaces to communicate with drivers in various tasks. Examples of such technology applications include:

a. *Autonomous Cruise Control (ACC)*: This is an intelligent driver assistance technology that regulates the speed of a vehicle with respect to vehicles running in front. The autonomous control systems hold the speed at constant level, and also slow down when traffic is sensed near the driver's vehicle, thus maintaining a safe distance to avoid mishaps.
b. *Lane Departure Warning Systems*: Lane departure warning systems are based on sensors equipped with laser vision and video mounted on both sides of the vehicle. Designed to emit warning signals to alert drivers when they leave their current lane, these systems minimize the potential risk of damage caused by accidents due to drowsiness and driving errors. Some vehicles combine hand gesture recognition–based media systems with lane-keeping systems to reduce the gaze distraction during driving.

Robotic assistance control systems (Figure 6.12) for driver comfort and safety are in great demand as they reduce the rate of fatalities caused due to distraction by mobile phone, text messaging on the way, drunkenness, and drowsing while driving a car. Having said this, it cannot be assured that everyone makes use of these technologies, but it can be surmised that using ADAS and sixth sense robotic technology will prevent casualties to a certain extent. Usability of a human-machine interaction system can be defined through five major quality components as depicted in Figure 6.13.

Learning Ability: A smart gesture recognition–based system for in-vehicle applications should be able to integrate natural hand gestures in such a way that gestures can be perceived with little or no cerebral load to perform driving activity and should be able to map generic system functions to easy and unique trained gestures.

Memory: A system should be designed in a user-friendly manner and should provide easily learnable gestures for basic functions, thus allowing the user to quickly memorize the important functions using small gesture symbols.

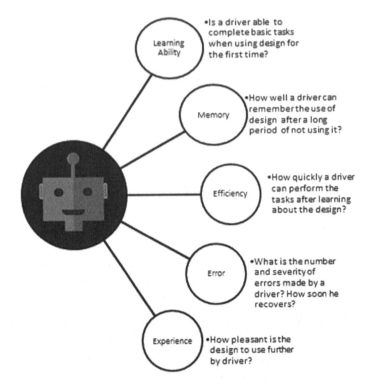

FIGURE 6.13: Quality components of an HMI assistance system [24].

Efficiency: When coupled with free hand gestures, the HCI system should be able to provide a faster channel of interaction and introduction of the new technology, while interfacing should not impact the efficiency of the driving task.

Error: The design and development of the system should procure minimal errors in recognizing gestures for interaction. The framework of touch gestures and free hand gestures in the air should be mapped precisely using computing algorithms or smart devices that could deal with different finger alignments for gesture interaction.

Experience: A realistic experience plays an important role in the design of human-robot interaction or human-machine interaction systems. The system should be provided with intuitive and interactive features which satisfy the user in terms of its interfacing and communication on the first go. Classification of free hand movements into small training functions provides for a high-quality user experience with ergonomic advantages.

6.11 REAL-TIME SIXTH SENSE GESTURE RECOGNITION SYSTEMS

In recent years, the gesture recognition, robotics, and sensory systems market has witnessed tremendous increase in demand with continuous advancements. A major driver for this growth is the increasing shift to sixth sense technology for connectivity in the consumer electronics, communication, healthcare, manufacturing, defense, retail, and automotive industries. The gaming and entertainment industry is also benefited by gesture recognition products that deploy augmented reality; the vision is to offer immersive user experience and low technical complexity across different applications.

Several advanced technologies are being used for gesture recognition, most of which are based on 2D or 3D sensors that work with the help of image-processing components like camera and integrated chip (IC). Of these, vision-based gesture techniques are used for electrical field sensing and security for commercial purposes. Other technologies are classified as device-based technologies that work on the principle of gesture interaction using gloves, markers, and motion trackers. The top-notch gesture recognition brands trending in the industry are as follows:

- *Nintendo Wii*: The Wii was released by Nintendo in 2006 as a seventh-generation home video game console. The Wii arrives to users with a Wii remote controller that can be used as a handheld pointer to detect the movement in three dimensions. The Wii console runs video games stored on Wii optical discs. It supports Internet communication services to receive updates while in standby mode. The Wii Mini, Nintendo's first major console redesign, was released first in 2012.
- *GestureTek Video Gesture Control*: As a pioneer and multiple patent-holder camera-enabled gesture-recognition technology, GestureTek offers a range of gesture-responsive technologies for real-time interaction between the user and the content that are used for presentation, education, hospitality, retail, and entertainment systems. In 1986, the company invented the domain of "applied computer vision" for human-computer interaction, and has been producing benchmark products based on interactive video gesture control technology (VGC) ever since. VGC technology allows users have easy control over multimedia information and manipulate immersive special effects in a virtual yet interactive digital world by making free hand or body movements.
- *Microsoft Gesture*: Microsoft developed Gesture as a cutting-edge, easy-to-use Software development kit (SDK) that creates a more natural and interactive experience by allowing the user to take control of communication through hand gestures. The Gesture SDK enables users to define their desired hand gestures using basic constraints built with natural language. Once a user's gesture is defined and registered in the software code, an alert message is generated whenever a user performs any hand gesture that allows him to choose the action in response. This technology can be used to control videos, play audio, bookmark online pages, and send emoticons as digital information.
- *Sony Music Gesture Control*: Sony LF-S50G gesture operations allow users to initiate Google Assistant gesture functions, play/pause a song, switch to previous and next song in the list, and adjust the volume with sensor and hand gestures. A user can perform touch-less hand gestures to control the functions of the wireless speaker by passing the hand over the gesture control system within the sensor area.

6.12 SEVENTH SENSE TECHNOLOGY – A NEW GESTURE INTERACTION STUDY

Seventh sense technology is a collaborative technology developed for human interaction with electronics and communication devices using natural hand gestures. This technology is a combination of advanced robotics and sixth sense capabilities that can control autonomous robots or microcontroller-based devices on demand by a human individual. Seventh sense enhances the capabilities of sixth sense applications with a cost efficient, intelligent, and human-friendly approach to connect with the digital world. The need for seventh sense technology lies in the few but major limitations of sixth sense such as the following:

- Sixth sense is a human-oriented gesture interaction technology and thus requires the presence of a human in the environment or space for system development.
- It doesn't support the control of robots or microcontroller-based devices, and only an authorized user can access the digital information.
- The process of gesture recognition needs the use of color markers to produce signals or information for human-machine interaction.

6.12.1 COLLABORATIVE ROBOTICS AND SENSOR TECHNOLOGY

The present-day design of autonomous robots is aimed at fulfilling a particular requirement or purpose. But with the advent of seventh sense technology, these robots and even microcontroller planted devices can achieve multipurpose objectives. Seventh sense technology outperforms sixth sense technology by allowing control of robots and such devices at any location using human natural hand gestures. Nevertheless, this technology shares the common underlying gesture recognition techniques as we discussed for sixth sense earlier.

The major role in the implementation of seventh sense technology is played by robots which perform actions based on the program flashed in the embedded microcontroller. A complete autonomous robot possesses the following capabilities:

1. Acquire information from surrounding environment using sensors
2. Charge battery using solar power without human assistance
3. Offer long life and better work accuracy
4. Adapt to new environment, gain new skills, and perform quickly

Seventh sense as a novel technology is the next-level approach toward physical world and digital world communication in the real environment without human intervention or the presence of an individual at all times. It provides the choice of customizing the virtual keyboard to equip devices with a greater number of actions desired.

6.12.2 HOW DOES SEVENTH SENSE TECHNOLOGY WORK?

Humans have the ability to make decisions based on the interpretation by their five biological senses. These decisions can be made using artificial neural networks

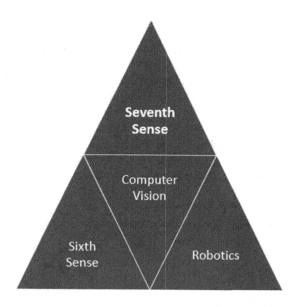

FIGURE 6.14:   Technologies related to seventh sense.

based on machine intelligence. Seventh sense implies a state of humans' ability to forecast the event and make changes to the things at the location where they are not actually present. This necessitates human-electronic interaction through robots or signaling devices with the incorporation of sixth sense technology. The most striking feature of the seventh sense model is the provision of using night vision cameras at low cost instead of sensors, which reduces the cost of implementation. Moreover, the introduction of seventh sense has given great attention to advanced research in the world of computer vision, robotics, and human-machine interaction technology.

Whereas sixth sense technology bridges the need for gesture-based communication to exchange information, seventh sense is assumed to be capable of controlling the robots or any microcontroller devices at distant locations by using natural hand gestures. This mainly applies to devices which have some type of microcontroller embedded inside. The modern automobile sector uses autonomous car-manufacturing robots that can perform tasks by themselves in controlled working environments. However, the movement of robots is restricted to a confined area in their workplace, which poses challenges in handling unpredictable potential hazards (Figure 6.14).

According to Sidharth Rajeev et al. [25], seventh sense technology uses the short-range IEEE 802.15.4 components for exchanging information. Using wireless transceivers, a seventh sense device can easily transmit sixth sense data to the target device placed remotely. It connects people to the digital world by giving a means of interaction with intangible information through a virtual keyboard. Human hand gestures are used to code instructions into electronic devices or robotic systems.

6.12.3 IMPLEMENTATION

Figure 6.15 illustrates seventh sense as a technology stack built upon the pillars of advanced sixth sense and robotics, which are backed by the instrumental foundation of computer vision. The robotics application of this technology is achieved by autonomous robots equipped with sensor modules for detection. These sensor modules send high pulse signals to direct the motion when an object's surface is detected, and send low pulse signals for direction when an object's edge is detected.

Seventh sense as a technology is operationalized with sixth sense (Section 6.3) using computer vision, gesture recognition, image processing, augmented intelligence, and neural network capabilities channelized on the wireless network.

Elements of Seventh Sense Communication

The design of a device based on seventh sense can be developed and implemented by application of the following elements:

1. Device Hardware
2. Programming Software
3. Gesture Recognition

Device Hardware

The device hardware consists of a camera, a projector, a mobile or PC, a microcontroller device, and a wireless transceiver.

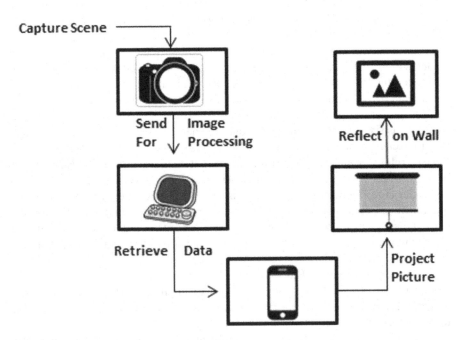

FIGURE 6.15: Block diagram of sixth sense implementation.

a) *Camera*: A camera is used to capture views of the scene in front of the user or object it is looking at. It collects the hand movements and different gestures made by the user.

b) *Projector*: A projector is used to augment the visualization of the object in context, projecting the information on a graphical user interface onto a target surface.

c) *Mobile or PC*: A mobile or PC acts as a computing device which performs suitable vision tasks and computation on the stream of media received from the camera. These can also be used to assist data communication with the projector.

d) *Microcontroller Device*: A microcontroller-based device is able to receive the series of data for communication from the mobile device. On accepting the information requests, this device performs operations led by instructions transmitted through the microcontroller. The device in control can be upgraded with a wireless camera to send and receive information from the surroundings, on the basis of which actions can be performed.

e) *Wireless Transceiver*: The seventh sense device prototype uses here IEEE 802.15.4, which is based on a ZigBee or GSM transceiver for communication within a larger area. For extended range, satellite-based transmitters and receivers can also be used.

Programming Software

Frequently used software and programming languages for image processing and visualization include MATLAB, Arduino, OpenCV, and Python, and Java and Embedded C, respectively.

Gesture Recognition

Gesture recognition in seventh sense is based on the fundamentals of sixth sense except for the use of a virtual keyboard instead of human intervention. Classification and recognition of gestures are achieved using either marker or free hand movement detection (Figure 6.16).

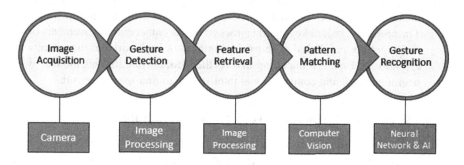

FIGURE 6.16:   Stages of gesture recognition process.

- **Image acquisition:**
  Images in the real-time environment are captured using a webcam or night vision camera. The captured image is sent to the software for processing, where the stream of scenes is converted into segments of image frame by frame.
- **Gesture detection:**
  This process is executed using color markers or bare hand movements depending upon the application. During this phase, image-processing algorithms and mathematical equations are used to filter background noise in order to enhance the detection of color points. For free hand gestures, Canny and SIFT algorithms have been used for image segmentation [25]. The gestural movements and every action performed by user are then stored using hardware.
- **Feature retrieval:**
  Smoothening of the image using threshold parameters is done for noise removal before analysis and segmentation of desired gestures. A feature extraction algorithm is then implemented to reproduce image edges and eventually detect the target features within the segmented frames.
- **Pattern matching:**
  This step of gesture recognition is executed following both marker and natural hand motion detection. Once the actual gestural segments with target features are obtained, these are compared with stored dataset of gesture templates using a pattern-matching algorithm including Euclidian distance and correlation-based techniques. In the final step, the designed system leverages the most accurately matched gestural feature to perform the corresponding action.
- **Gesture recognition:**
  Virtual keyboard technique has been widely used to secure and authenticate user access to a system. The most common examples include the profile authentication while using banking systems or secure payment portals. In seventh sense as a technology framework, a virtual keyboard system is used to enhance the gesture recognition process for remote communication. This technique doesn't limit the user to the number of gestures that can be made by fingers. Shuai Yang et al. [4] described the virtual keyboard operation. The use of a red marker on one finger is aimed at tracking the keys pressed by the user, which are dictated to the user with alert notifications. The projected virtual keyboard is tracked by the camera, and movements of the finger are monitored. The tracked images of keyboard and movements are stored for pattern matching in a further process that implies the use of neural network and computational intelligence to finalize the gestures.

6.13   SEVENTH SENSE AS NOVEL DESIGN PANOPLY

A robot working with seventh sense technology is designed to have three basic elements: sensor, microcontroller, and actuator. The sensor device traces the motion and path of the object. This information as input is sent to the microcontroller in the form of signals to decide the actions to be performed by the robot. The instructions are then passed on to actuators to perform movement of robotic arms corresponding to the gestural actions. As discussed earlier, a robotic system with seventh sense has a transceiver attached to pass the information signals with the help of a camera-assisted computing device. With an advanced sixth sense mechanism and robot-assisted development, seventh sense digital prototyping can lead to revolution in design for many applications like the following:

- Nano-robotics for healthcare
- Medicinal treatment for cancer
- Remote surveillance to deter trespassers
- Assistance during surgical operations
- Tracking and altering satellite orbit path
- Gestural action technique for disabled
- Remote in-vehicle driving assistance
- Gesture-based robotic control
- Gesture-based techniques for E-learning

6.14 SUMMARY

Sixth sense technology is a technology used for human-computer interaction using gestures. The hand gestures can be identified and recognized using computational intelligence. Sixth sense as a technology is driven by augmented reality and computer vision algorithms to state the target action. The latest use of sixth sense in robotic in-vehicle assistance has paved the path to further research on the industrial applications of gesture-based device control systems. Nevertheless, a robot with a sixth sense–enabled sensor can perform tasks which a human can do in real time. The advancement in sixth sense technology is the introduction of seventh sense, which eliminates the requirement for humans in all situations to train gestural actions. Seventh sense technology makes use of microcontrollers to govern the actions of robots at distant places by using cameras, wireless communication setups, and virtual keyboard systems. Seventh sense as a technology is comprised of sixth sense technology and advanced robotics working on the principles of computer vision and image processing to achieve multipurpose design goals. The interconnection of sixth sense with the concept of seventh sense technology is going to attract the attention of designers and practitioners toward a new space of research and development that will shape the future world of technology.

REFERENCES

1. Amit Konar, Sriparna Saha. "Gesture Recognition: Principles, Techniques and Applications", *Studies in Computational Intelligence*. Springer 724 (2018).
2. Yanxi Zhang, Yongjian Liang, Daowen Wu, Qiang Wen, Hengyuan Yang. "Gesture Acquisition and Tracking with Kinect under Complex Background", *Applied Mechanics and Materials 511–544*. Switzerland, Trans Tech Publications (2014).

3. Amal Kumar Das, Vijaya Laxmi, Sanjeev Kumar. "Hand Gesture Recognition and Classification Technique in Real-Time", *Proceedings of IEEE International Conference on Vision Towards Emerging Trends in Communication and Networking (ViTECoN)*. Vellore (2019): 1–5.
4. Shuai Yang. *Robust Human Computer Interaction Using Dynamic Hand Gesture Recognition*, Research Thesis, University of Wollongong (2016).
5. P. Premaratne, S. Yang, Z. Zhou, N. Bandara. "Dynamic Hand Gesture Recognition Framework", *Intelligent Computing Methodologies. ICIC 2014. Lecture Notes in, Computer Science 8589*. Springer, Cham (2014).
6. Roland Aigner, Daniel Wigdor, Hrvoje Benko, Michael Haller, David Lindbauer, Alexandra Ion, Shengdong Zhao, Jeffrey Tzu Kwan Valino Koh. "Understanding Mid-Air Hand Gestures: A Study of Human Preferences in Usage of Gesture Types for HCI", *Microsoft Research Technical Report MSR-TR-2012-111*. Microsoft Corporation (2012).
7. C. Cadoz. *Les réalités virtuelles*. Flammarion, Paris (1994), ISBN: 2-08-035142-7.
8. Enrique Coronado, Jessica Villalobos, Barbara Bruno, Fulvio Mastrogiovanni. "Gesture-Based Robot Control: Design Challenges and Evaluation with Humans", *IEEE International Conference on Robotics and Automation (ICRA)*, Singapore (2017).
9. M. Karam, M.C. Schraefel. "A Taxonomy of Gestures in Human Computer Interactions", *Technical Report ECSTR-IAM05-009, Electronics and Computer Science*. University of Southampton (2005).
10. J. Rodrigues Joao, Pedro Cardoso, Jânio Monteiro, Mauro Figueiredo. *Handbook of Research on Human-Computer Interfaces, Developments, and Applications, Advances in Human and Social Aspects of Technology Series*. IGI Global (2016).
11. Santiago Riofrio, David Pozo, Jorge Rosero, Juan Vasquez. "Gesture Recognition Using Dynamic Time Warping and Kinect: A Practical Approach", *IEEE International Conference on Information Systems and Computer Science (INCISCOS)* (2017): 302–308.
12. Md. Shoaibuddin Madni, Ravindra N. Rathod. "Color Segmentation for Sixth Sense Device", *Bonfring International Journal of Research in Communication Engineering 6*, Special Issue (November 2016).
13. Andrew D. Wilson, Aaron F. Bobick. "Parametric Hidden Markov Models for Gesture Recognition", *IEEE Transactions on Pattern Analysis and Machine Intelligence* 21(9) (1999):886–900.
14. Nhan Nguyen-Duc-Thanh, Sungyoung Lee, Donghan Kim. "Two-Stage Hidden Markov Model in Gesture Recognition for Human Robot Interaction", *International Journal of Advanced Robotic Systems, 9/2*. SAGE Publishing (2012).
15. Byung-Woo Min, Ho-Sub Yoon, Jung Soh, Yun-Mo Yang, Toskiaki Ejima. "Hand Gesture Recognition Using Hidden Markov Models", *IEEE International Conference on Systems, Man, and Cybernetics. Computational Cybernetics and Simulation* 5 (1997): 4232–4235.
16. Hugo Jair Escalante, Eduardo F. Morales, L. Enrique Sucar. "A Naïve Bayes Baseline for Early Gesture Recognition", *Pattern Recognition Letters*. Elsevier Science (2016).
17. Andrew David Wilson. "Adaptive Models for the Recognition of Human Gesture", PhD thesis, MIT Program in Arts and Sciences (2000).
18. Hector Hugo Avilts-Aniaga, Luis Enrique Sucart, Carlos Eduardo Mendozaz, Blanca Vargas. "Visual Recognition of Gestures Using Dynamic Naive Bayesian Classifiers", *Proceedings of the 2003 IEEE International Workshop on Robot and Human Interactive Communication Millbrae*. California (2003): 133–138.
19. Premangshu Chanda, Pallab Kanti Mukherjee, Subrata Modak, Asoke Nath. "Gesture Controlled Robot Using Arduino and Android", *International Journal of Advanced Research in Computer Science and Software Engineering 6*, 6 (2016), ISSN: 2277 128X.

20. M.A. Goodrich, A.C. Schultz. "Human-Robot Interaction: A Survey", *Foundations and Trends in Human-Computer Interaction* 1(3) (2007): 203–275.
21. D.A. Norman. *The Design of Everyday Things*. MIT Press, Cambridge (1988).
22. P. Premaratne, Q. Nguyen. "Consumer Electronics Control System Based on Hand Gesture Moment Invariants", *Computer Vision, IET* 1 (2007): 35–41.
23. M. Thomas Kopinski. "Neural Learning Methods for Human-Computer Interaction", *Machine Learning [cs.LG]*. Université Paris-Saclay (2016).
24. Jakob Nielsen. *Usability Engineering*. Elsevier, Cambridge (1994).
25. Rajeev Sidharth. "Seventh Sense Technology", *Proceedings of the IEEE UP Section Conference on Electrical Computer and Electronics (UPCON)* (2015): 1–10.

Index

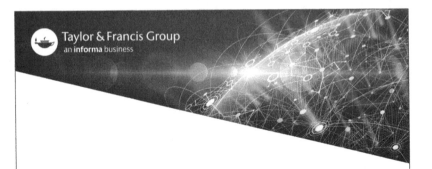

Printed in the United States
By Bookmasters